THE FIFTH GENERATION (5G) OF WIRELESS COMMUNICATION

Edited by **Ahmed Kishk**

The Fifth Generation (5G) of Wireless Communication
http://dx.doi.org/10.5772/intechopen.71356
Edited by Ahmed Kishk

Contributors

Mark Leeson, Matthew Higgins, Ali H. Alqahtani, Kioumars Pedram, Mohsen Karamirad, Negin Pouyanfar, Wen Xu, Armin Dammann, Tobias Laas

Notice

Statements and opinions expressed in the chapters are these of the individual contributors and not necessarily those of the editors or publisher. No responsibility is accepted for the accuracy of information contained in the published chapters. The publisher assumes no responsibility for any damage or injury to persons or property arising out of the use of any materials, instructions, methods or ideas contained in the book.

First published in London, United Kingdom, 2019 by IntechOpen
IntechOpen is the global imprint of INTECHOPEN LIMITED, registered in England and Wales, registration number: 11086078, The Shard, 25th floor, 32 London Bridge Street
London, SE19SG – United Kingdom
Printed in Croatia

British Library Cataloguing-in-Publication Data
A catalogue record for this book is available from the British Library

Additional hard copies can be obtained from orders@intechopen.com

The Fifth Generation (5G) of Wireless Communication, Edited by Ahmed Kishk
p. cm.
Print ISBN 978-1-78985-769-6
Online ISBN 978-1-78985-770-2

We are IntechOpen,
the world's leading publisher of
Open Access books
Built by scientists, for scientists

4,000+
Open access books available

116,000+
International authors and editors

120M+
Downloads

Our authors are among the

151
Countries delivered to

Top 1%
most cited scientists

12.2%
Contributors from top 500 universities

Interested in publishing with us?
Contact book.department@intechopen.com

Numbers displayed above are based on latest data collected.
For more information visit www.intechopen.com

Meet the editor

Dr. Ahmed Kishk is at Concordia University, Montréal, Québec, Canada (since 2011) as Tier 1 Canada Research Chair in Advanced Antenna Systems. His research interest includes the areas of design of Dielectric resonator antennas, microstrip antennas, small antennas, microwave sensors, RFID antennas for readers and tags, Multi-function antennas, microwave circuits, EBG, artificial magnetic conductors, soft and hard surfaces, phased array antennas, and computer aided design for antennas; Design of millimeter frequency antennas; and Feeds for Parabolic reflectors. He has published over 290-refereed Journal articles and 450 conference papers. He is a coauthor of four books and several book chapters and the editor of three books. He offered several short courses in international conferences. Kishk is a Fellow of IEEE since 1998.

Contents

Preface

This Edited Volume is a collection of reviewed and relevant research chapters, concerning the developments within the fifth generation (5G) of wireless communication. The book includes scholarly contributions by various authors and is edited by an expert pertinent to Electrical and Electronic Engineering. Each contribution comes as a separate chapter complete in itself but directly related to the book's topics and objectives.

The book is compiled of the following chapters: *Introductory Chapter: The Future of Mobile Communications, Optical Wireless and Millimeter Waves for 5G Access Networks, Rateless Space-Time Block Codes for 5G Wireless Communication Systems, Evolution and Move toward Fifth-Generation Antenna* and *Where are the Things of the Internet? Precise Time of Arrival Estimation for IoT Positioning.*

The target audience comprises scholars and specialists in the field.

IntechOpen

Chapter 1

Introductory Chapter: The Future of Mobile Communications

Mark Stephen Leeson

Additional information is available at the end of the chapter

http://dx.doi.org/10.5772/intechopen.84503

1. Introduction

The availability of mobile communication links has become an integral part of modern societies even, or perhaps especially, when other infrastructure is lacking [1]. Moreover, projections of mobile data traffic growth indicate that there will be five times as much traffic by 2030 even in a low-growth scenario, and it is quite likely that data traffic will be ten times greater than today [2]; to address this expansion requires increased system capacity. The now ubiquitous wireless systems in modern life operate using carrier frequencies below 6 GHz. The current, fourth-generation (4G) systems have been in use for over a decade and are reaching their limits given the increased data traffic demands. Thus, it is time for fifth-generation (5G) mobile communication networks to be introduced.

Previous cellular mobile networks were developed primarily to optimise voice or video streams and supported other services such as Web browsing as additional features. However, the emergence of many new applications, typified by autonomous vehicles, smart cities, virtual reality and ubiquitous remote control to name a few has placed extra requirements on the mobile infrastructure. As a consequence, 5G is a paradigm shift from the cellular network of 4G to a 'network of everything', connecting people and machines in a service-based architecture. The International Telecommunication Union Radiocommunications (ITU-R) Standardization Sector has thus specified a number of goals for 5G in the standard International Mobile Telecommunications 2020 (IMT-2020) [3]. The aims include a 20 Gbps peak data rate and 100 Mbps user experienced data rate coupled with high density and low latency and the accommodation of high-speed mobility. Progress towards these design goals has been led by the Third Generation Partnership Project (3GPP) [4], an umbrella standards body that has defined the 5G standard, often referred to as new radio (NR). This standard is currently undergoing development of its 16th release following the completion of release 15 in 2018. The latter concentrated on enhanced mobile broadband to provide high data rates for consumer applications such as video on demand. The former, currently underway, considers ultra-reliable low-latency communications for uses such as remote surgery and the need to support a large number of devices for machine to machine communications in the Internet of Things (IoT) [5].

The targets of 5G are challenging and require the development of technology in many areas. In this book, some of the aspects of 5G are addressed to accommodate the increased services, data rates and reliability promised by this next generation of mobile communications.

Physical layer developments are the focus of Leeson and Higgins, who consider the movement towards new wavelength ranges for future communication systems driven by increasing bandwidth demands as 5G comes to fruition. In particular, using two topical technologies offer substantial transmission bandwidth via high carrier frequencies, namely, optical wireless and millimetre-wave transmission. Following an introduction to the relevant portion of the electromagnetic (EM) spectrum for these technologies, a short history of their development is given. Significantly, a performance comparison for 5G outdoor point-to-point and indoor hotspots is presented. This indicates that optical wireless is the better option for distances up to approximately 50 m outdoors and an indoor 10 m hotspot radius. Time will tell if these two approaches will become established as 5G implementation becomes a reality, but they offer a possible future for future wireless data demands using untapped and unlicensed bandwidth in relatively free portions of the EM spectrum.

The introduction of 5G systems introduces requirements for a range of antennas that are discussed by Pedram et al. in their chapter. Novel approaches are necessary to address 5G antenna provision for a collection of frequency bands, usually in the GHz range. Although it is true that most transmission and reception constraints due to antenna design have been removed from existing systems, this is not true for the additional demands of 5G. The chapter is organised into three subsections that cover antennas below 15 GHz, antennas operating between 15 and 30 GHz and higher frequency antennas. New antenna technologies such as metamaterial structures, substrate integrated waveguide structures and microstrip antennas with various feeding networks are considered as appropriate in the 5G frequency ranges stated. The trade-offs needed when choosing the frequency of operation, such as size and bandwidth, are also discussed.

Delivery of high-quality communications also requires the use of modern coding methods without which wireless communication would be impossible at the rates needed for 5G. Thus, Alqahtani provides a chapter that presents a rateless space-time block code (RSTBC) for massive multiple-input multiple-output (MIMO) wireless communication systems. The principles of rateless coding compared to the fixed-rate channel codes are discussed, followed by presentation of a literature review of rateless codes (RCs). Furthermore, the chapter illustrates the basis of RSTBC deployments in massive MIMO transmissions over lossy wireless channels. In such channels, data may be lost or are not decodable at the receiver end due to a variety of factors such as channel losses or pilot contamination. Massive MIMO is a breakthrough wireless transmission technique proposed for future wireless standards due to its spectrum and energy efficiencies. The chapter thus presents details of a technology that is central to the implementation of 5G.

With the introduction of communication between machines in 5G systems, Xu et al. provide a chapter that considers the accurate positioning of the diverse range of 'things' that may be connected to form the Internet of Things (IoT) in 5G communication networks. With the expansion of mobility provided by objects such as automobiles and drones, knowledge of the location of these entities is taking on a growing importance for future mobile networks. The chapter provides an overview of the mobile radio positioning techniques that have been used by the various generations of mobile communication systems. This gives the most detail concerning 4G positioning employing user equipment and base stations, since this will most likely be extended for 5G. The use of higher transmission frequencies in 5G will enable more accurate measurement of time of arrival to determine distances. In addition, the proliferation of connected devices will enable more cooperative working and further contribute to high-accuracy positioning.

So, this collection of leading research chapters is intended to provide information for researchers and practitioners in communication systems. It may be used as a reference for the latest developments in 5G, and as a pointer to developing topics for advanced students looking for insight into topics for dissertations and future research.

Author details

Mark Stephen Leeson

Address all correspondence to: mark.leeson@warwick.ac.uk

School of Engineering, University of Warwick, Coventry, UK

References

[1] Anstey Watkins JOT, Goudge J, Gómez-Olivéc FX, Griffiths F. Mobile phone use among patients and health workers to enhance primary healthcare: A qualitative study in rural South Africa. Social Science & Medicine. 2018;**198**:139-147. DOI: 10.1016/j.socscimed.2018.01.011

[2] Oughton E, Frias Z, Russell T, Sickerd D, Cleevely DD. Towards 5G: Scenario-based assessment of the future supply and demand for mobile telecommunications infrastructure. Technological Forecasting and Social Change. 2018;**133**:141-155. DOI: 10.1016/j.techfore. 2018.03.016

[3] Lien SY, Shieh SL, Huang Y, Su B, Hsu YL, We HY. 5G new radio: Waveform, frame structure, multiple access, and initial access. IEEE Communications Magazine. 2017;**55**:64-71. DOI: 10.1109/MCOM.2017.1601107

[4] 3GPP The Mobile Broadband Standard [Internet]. 2019. Available from: http://www. 3gpp.org [Accessed: January 20, 2019]

[5] Gazis V. A Survey of Standards for Mvachine-to-Machine and the Internet of Things. IEEE Communications Surveys & Tutorials. 2016;**19**:482-511. DOI: 10.1109/COMST.2016.2592948

Chapter 2

Optical Wireless and Millimeter Waves for 5G Access Networks

Mark Stephen Leeson and Matthew David Higgins

Additional information is available at the end of the chapter

http://dx.doi.org/10.5772/intechopen.77336

Abstract

Growing bandwidth demands are driving the search for increased network capacity leading to the exploration of new wavelength ranges for future communication systems. Therefore, we consider two technologies that offer increased transmission bandwiths by virtue of their high carrier frequencies, namely optical wireless and millimeter-wave transmission. After highlighting the relevant electromagnetic (EM) spectrum region, we briefly describe the applications and properties of each approach coupled with a short history of their development. This is followed by a performance comparison in two possible 5G links: outdoor point-to-point and indoor hotspots. We find that in both cases, there are regions where optical wireless communications (OWC) are better, but others where millimeter waves are to be preferred. Specifically, the former outperforms the latter over distances up to approximately 50 meters outdoors and a 10-meter hotspot radius indoors.

Keywords: optical wireless communications (OWC), visible light communications (VLC), free space optics, infrared (IR) communications, millimeter-wave communications, 5G access

1. Introduction

Mobile data traffic is projected to increase by several orders of magnitude by the year 2030 [1] and to address this expansion requires increased system capacity. The now ubiquitous wireless systems in modern life operate using carrier frequencies below 6 GHz. The next generation of provision must be designed to meet future wireless data demands, and so the search for further regions of the electromagnetic (EM) spectrum with untapped bandwidth has continued with renewed vigor in recent years. This has prompted research interest in underutilized

Wavelength (m) 10^{-1} 10^{-3} 10^{-5} 10^{-7} 10^{-9}

10 mm   1 mm   750 nm 390 nm 10 nm

30 GHz   300 GHz   300 THz

Frequency (Hz) 10^{9} 10^{11} 10^{13} 10^{15} 10^{17} 10^{19}

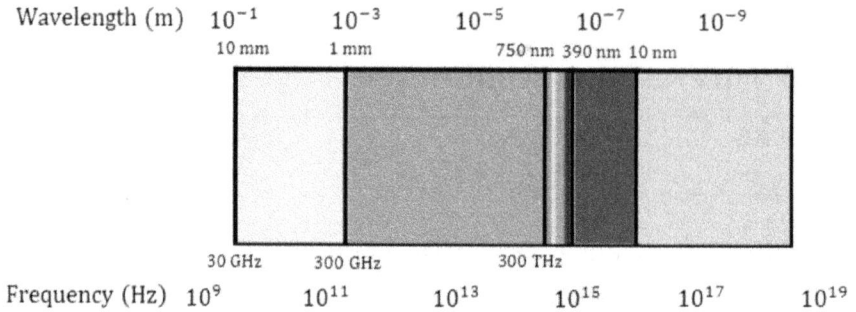

Figure 1. The relevant portion of the EM spectrum.

higher frequency bands, with millimeter [2] and optical [3] waves being prime areas of interest. **Figure 1** below shows a portion of the EM spectrum that includes the ranges for optical and millimeter-wave communication. The former includes a large span of frequencies that encompasses the infrared (IR) region through visible light and into the ultraviolet (UV) range. The latter includes wavelengths from 1 to 10 mm, equating to 300 GHz and 30 GHz, respectively.

As may be seen from **Figure 1**, millimeter-wave and optical frequencies live side-by-side in the EM spectrum and share the characteristics of propagation but are often seen as disparate entities. In many ways, it would seem sensible to classify optical communications as nanometer wave communications, but this has not been the case to date. However, both spectral regions offer significantly increased transmission bandwidths by virtue of their increased carrier frequencies, and this is the reason for their entry into the 5G arena. Nevertheless, the realization of the potential of both optical wireless communications (OWC) and millimeter waves requires solutions to the significant challenges that arise from the utilization of higher frequency carriers. Transmissions using both OWC and millimeter waves occur in relatively demanding propagation conditions. Both systems suffer from increased path loss, extra channel losses, and potentially useful technology that is not as established as that commonly deployed at present [4, 5].

A notable difference is that OWC has leveraged component advances from fiber systems [6], whereas the technological advances needed to make millimeter-wave radio cost-effective are more recent [7]. The rest of this chapter presents brief reviews of OWC and millimeter-wave systems and subsequently compares their performance in likely 5G scenarios.

2. Optical wireless

The use of optical carriers in free space is a technology that combines the mobility of radio frequency (RF) wireless communications with the high-potential bandwidth of optical communications. Moreover, the optical spectrum is not subject to license fees with easy spectrum reuse since light beams cannot penetrate walls. The major design challenge for OWC is to achieve a sufficiently high signal-to-noise ratio (SNR) at useful data rates given that the transmitter (TX) power is limited by eye safety considerations [8].

2.1. Application areas and properties

OWC systems can be broadly classified into categories, namely indoor and outdoor. Within both of these, a large number of possible operating modes have been demonstrated that may be grouped using the sub-categories shown in **Figure 2**. A good overview of the classification of optical wireless systems has recently been given by Son and Mao [9], to which the reader is referred for more details, so here we provide only an overview.

The two fundamental designs for indoor OWC are directed line-of-sight (LOS) and diffuse links, illustrated in **Figure 3**. In the former, a narrow beam TX sends light to a narrow field of view (FOV) receiver (RX) over the LOS. Such a link thus experiences minimal impacts from the multipath dispersion, noise, and path loss. Although the data rate is limited by the power budget allowed, such LOS links are very suitable for high-speed hotspots. The idea of tracking a user to support mobility with a TX and RX array [10] was demonstrated during the early revival of OWC [11], and this concept is likely to be employed with the light emitting diode (LED) lighting discussed later in this chapter.

A diffuse link relies on a wide-beam TX and a wide FOV RX and mirrors the operation of current WiFi systems by scattering an optical beam from surfaces within a room. Although this offers the advantage of more than one path, the differing path lengths produce a multipath dispersion effect that limits the achievable bit rates [12]. Despite innovations such as quasi-diffuse systems [13] and alternative modulation methods [14], the increased path loss produces

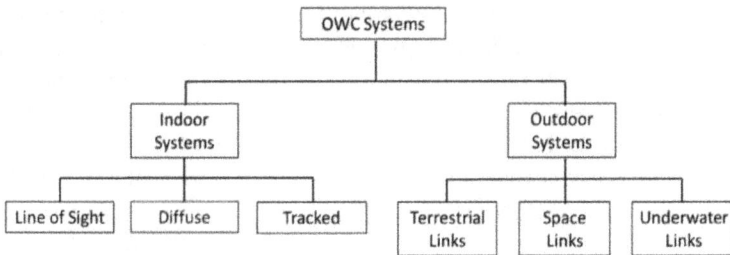

Figure 2. Simplified taxonomy of OWC.

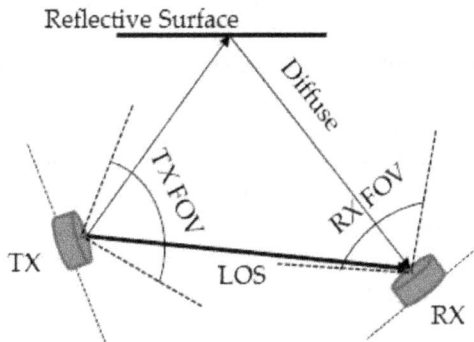

Figure 3. Schematic view of LOS and diffuse OWC links.

a need for higher power levels meaning that the diffuse systems are not competitive with RF solutions and unlikely to be employed in 5G systems.

Outdoor OWC is usually referred to as free space optical (FSO) communication and makes use of LOS links as the only feasible option given the path loss. We can consider FSO systems by means of their distance from the center of the earth. First, there are OWC satellite networks that can cover large portions of the globe [15]. Second, terrestrial FSO links [16] are usually established between buildings. Finally, there is the rapidly developing area of underwater OWC [17] where the incumbent acoustic technology is extremely limited in its bit rate, and OWC offers enhanced performance. Of these, we concentrate on terrestrial systems here since they are closest to the interface with 5G. We recognize that the satellites may be needed in the overall 5G landscape, but many of the issues will be similar (particularly in the area of ground to satellite communications), whereas underwater OWC is an important but distinct area with different propagation conditions.

2.2. Brief history

Communication using light through the air has a long history, beginning with the employment of reflected sunlight and smoke signals by ancient civilizations [18]. However, the birth of what may be regarded as a "modern" system dates from Bell's 200-meter photophone in 1880 [19]. With respect to FSO, the invention of the laser in the 1960s provided the means for a range of FSO applications [20]. Optical fiber developments in the next two decades made these preferable for long-distance optical transmission. Continued military and space work provided the basis for commercial FSO [21]. The 1990s saw growth in the civilian usage of FSO links driven by increasing data rates and high-quality connectivity requirements. FSO deployment rather than fiber offers a cheaper and quicker way of providing customer bandwidth, and may also be employed in disaster scenarios [22]. Substantial research efforts to improve FSO system performance in adverse atmospheric conditions mean that multi-gigabit rates are possible in the presence of turbulence [23].

Modern developments in indoor systems, commonly known as OWC, began with the seminal work of Gfeller and Bapst [24] in 1979. This considered an IR system based on a diffuse link using a wavelength of approximately 950 nm. The data rate achieved was just 125 kbps, but the work prompted the development of IR OWC systems through the 1980s and 1990s. Thus, LOS bit rates reached 155 Mbps [25], and diffuse systems achieved 70 Mbps [26]. This period of substantial development for OWC using wavelengths between 780 and 950 nm was driven largely by the ready availability of inexpensive optical sources and the coincidence of the peak sensitivity of mass-produced photodiodes. To maintain the essential simplicity and low-cost of OWC, most systems employed intensity modulation with direct detection (IM/DD), but many modulation schemes were investigated [27]. OWC did not achieve mass market status during this period but progress in IR systems has nevertheless continued with recent results demonstrating multi-gigabit performance [28] and localization and tracking [10]. However, it has been the developments in the visible light range that have really brought OWC to the fore as an option for integration with 5G. The development of solid-state lighting has led to the emergence of visible light communications (VLC) [29]. This approach makes use of the communication potential of the lighting system that is offered once white LEDs (WLEDs) are installed. VLC in its modern form was initiated primarily by work at Keio University in Japan during the early part of the millennium [30]. Over the next period of years, the increasing research interest in the field led to the creation of the Institute of Electrical and Electronics Engineers (IEEE) 802.15.7 VLC task group, which has standardized physical (PHY) and medium access control (MAC) layers, and characterized these for short-range data transmission [31].

Advances have continued thanks to the orthogonal frequency division multiplexing (OFDM) scheme where parallel orthogonal subcarriers are used to achieve high data rates [6]. Direct current biased OFDM (DCO-OFDM) enabled the demonstration of a data rate of in excess of 3 Gbps using a commercial LED [32]. In recent years, the term light fidelity (LiFi) has been introduced [33] to encompass the aspect of mobility envisaged in the latest systems compared to the original fixed point-to-point concept of VLC [34]. Development of the technology continues with the investigation, *inter alia*, of modulation schemes, modeling, and applications [35].

The twentieth century also saw the use of UV light for communications, making use of the wavelength range 200–280 nm where there is very low background noise due to strong atmospheric absorption, so UV offers non-LOS (NLOS) secure communication [36]. During the 2000s, work at Massachusetts Institute of Technology (MIT) Lincoln laboratory replaced bulky, slow gas discharge lamps with UV LEDs [37] but the employment of photomultiplier tubes (PMTs) until relatively recently [38] has been a weakness of UV and OWC. The modern alternative of an avalanche photodiode (APD) adds complexity, and although UV systems system work continues, it is difficult to envisage 5G UV deployment.

3. Millimeter waves

There has been considerable interest in millimeter waves, particularly around 28-, 38-, 60 GHz, and the E-band (71–76 GHz and 81–86 GHz) [39]. The progress in complementary metal-oxide-semiconductor (CMOS) RF integrated circuits for 60 GHz systems [40] offers future prospects for products. Millimeter-wave communications have similarities with OWC since the higher carrier frequency suffers a high propagation loss, and is more sensitive to blockage than existing RF systems.

3.1. Application areas and properties

The applications of millimeter waves may be broadly classified as shown in **Figure 4**, in which systems where the millimeter waves convey information are distinguished from those where they serve another purpose. In the first category, outdoor cellular transmission for 5G has attracted substantial attention. The reduced propagation range of millimeter waves means that to achieve the increased bandwidth and throughput, smaller outdoor transmission ranges will be used [2] as illustrated later schematically in **Figure 6**. As will be discussed further when millimeter waves are compared with OWC, both technologies can form the basis for high-speed WiFi links to offer increased bandwidth and thus greater capacity [41]. Millimeter waves

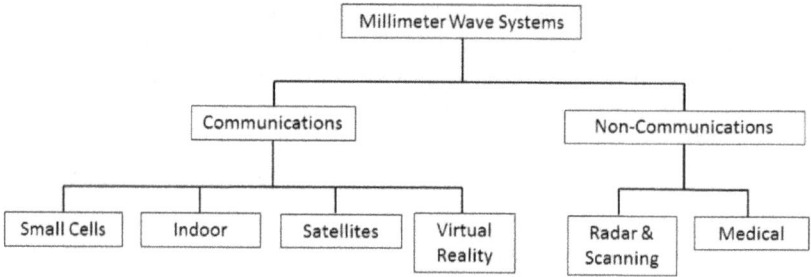

Figure 4. Major application categories of millimeter waves.

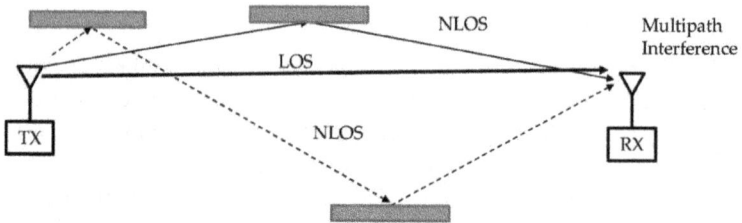

Figure 5. Multiple propagation paths in millimeter-wave transmission.

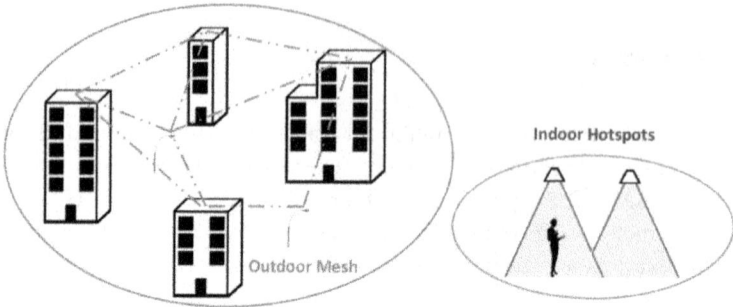

Figure 6. Comparison scenarios.

are ideal for satellite communications because of their significant bandwidth [42] and for high-speed transmission of video and audio for virtual reality (VR) applications [43]. Millimeter-wave radar has also been widely applied, most latterly in autonomous vehicles [44] and contraband detection [45]. Medical applications include cancer treatment by the use of immune system therapy [46].

A schematic representation of a millimeter-wave link is shown in **Figure 5**, which illustrates two possible reflected paths by which the transmitted waves may reach the RX. In contrast to OWC, these NLOS links cause multipath interference fading effects at the scale of a wavelength [47] because constructive and destructive effects can occur. This not seen in OWC because photo-diodes are typically many thousands of wavelengths across providing spatial diversity that prevents the fading although not pulse dispersion. The advances in beamforming that are outlined in the next section have enabled NLOS transmission to be implemented [48].

3.2. Brief history

Transmissions using millimeter-wave carriers have a long history but millimeter-wave mobile communications arose in the 1980s [49]. There was then a substantial gap in millimeter-wave communications research until the release of the unlicensed band near to 60 GHz [50]. This led to the development of short-range Gbps point-to-point links and wireless network standards [51]. As with OWC, military applications were developed over a similar time period [52], recognizing the potential benefits of increased bandwidth, sophisticated antennas, greater directionality, and reduced size compared to traditional microwave links.

Two key technological developments have enabled 60 GHz systems to become a reality, namely high-speed integrated circuits [40] and the use of multiple antennas [53]. With respect to the first of these, the work that has led to the emergence of low-cost millimeter-wave circuitry began in the 1990s [54], employing III-V semiconductor compounds that were hard to integrate with digital circuitry. Some of the earliest work using 60GHz CMOS appeared in the 2000s [55] progressing to the low-power implementation of Alldred et al. [56]. Strides in integration later resulted in chips of only a few square millimeters in area, including an antenna and with power consumption under 100 mW [57]. Gbps speeds over 2 meters were also demonstrated using integrated architectures [58]. The technology is now at a point where sophisticated modulation techniques can be employed at multi-gigabit rates for the latest IEEE standards using low supply voltages and small chip areas [59].

With respect to multiple antennas, the formation of arrays added considerable design flexibil-ity to millimeter-wave systems. The fundamental driver of the technology was, and remains, the need to compensate for the large propagation loss incurred by this wavelength range. The utilization of highly directional antennas provides the necessary gain and the ability to imple-ment beamforming that will be discussed below. The small wavelength is an advantage since the antenna size is half a wavelength as is the antenna separation permitting many antennas to be fitted into a relatively small space (several per square centimeter). Although the idea of using antenna arrays and adapting their beam patterns has its origins in radar systems and has existed for some time [60], interest in the concept for wireless communications started in the 1990s. Beamforming describes a signal processing technique used to achieve directional trans-mission or reception of a wireless signal. The selectivity in the antenna patterns is implemented by adapting the beams in an operation that may be seen as linear filtering in the spatial domain. The phases and amplitudes of the signals may be controlled to produce maxima or minima in desired and undesired communication directions, respectively [61]. Beamforming can be implemented in the digital baseband, the analog baseband or the RF front-end [53], and each of these has its own particular design features [62]. The filter weights employed to drive the antennas may be fixed, but a more flexible method (particularly for mobile communications) is adaptive or smart beamforming because it can adapt the RF radiation pattern in real-time to accommodate changing transmission conditions [63].

4. Performance comparison

Based on the properties of the both OWC and millimeter-wave transmission, we now focus on a comparison of their performance in the two of the most likely application scenarios shown in **Figure 6**. Firstly, in the outdoor arena point-to-point links may be used to establish a mesh network using individual "point-and-shoot" links. Secondly, high-speed indoor connectivity may be offered using hotspots.

4.1. Methodology

We adopt a simplified approach so that the broad sweep of the performance of the technologies is captured. It is inevitable that some subtleties will be overlooked (such as shadowing and the mechanism to provide a primarily LOS path), but the analysis enables a meaningful indication of the relative performance of the two PHY layers. We now describe the simplified link budget models of the optical and millimeter-wave channels. These are along the lines originally described in [64] for a previous technology generation and inspired by the analysis of Wolf and Kress [65]. We consider both outdoor and indoor LOS scenarios.

4.1.1. Point and shoot

Figure 7 shows the geometry of the outdoor application where both transceivers have emission and FOV half angles of θ_h for simplicity and are not necessarily aligned to be facing each other.

4.1.1.1. Optical wireless

Applying the Friis formula [66] to an optical link produces unrealistic results because it applies to narrow, diffraction limited beams that require very precise alignment. Therefore, we adopt the more customary approach for OWC [67] where a link is considered that launches a beam with half-angle θ_h that evenly illuminates the area within the emitter cone. The intensity profile is assumed to be a constant value of I_0 over an angle $[0, \theta_h]$. This profile is beneficial so

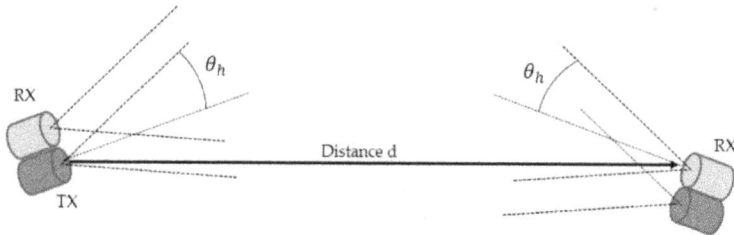

Figure 7. Point and shoot configuration.

that there is no off-axis fall-off, and may be obtained using a holographic diffuser [68]. Considering the solid angle of the cone results, for an emitter source power P_S, in:

$$I_0 = \frac{P_S}{2\pi r^2 (1 - \cos\theta_h)} \tag{1}$$

The RX has a collection area, and in the worst case, this is orientated at an angle θ_h, meaning the received optical power P_O is:

$$P_O = \frac{P_S}{2\pi r^2 (1 - \cos\theta_h)} A_{coll} \cos\theta_h \tag{2}$$

The photocurrent resulting from the received optical power is $i_O = RP_O$, for a photodiode responsivity R. Therefore, the electrical power S, delivered to a load R_L is:

$$S = I_O^2 R_L \tag{3}$$

The noise at the RX comprises shot noise from the signal, shot noise from any DC photocurrent caused by ambient light and amplifier noise. Thus, the noise power delivered to the load over an amplifier bandwidth Δf is:

$$N = \left(2q(RP_O + i_{Amb})\Delta f + i_{Amp}^2 \Delta f\right) R_L \tag{4}$$

where i_{Amb} is the photocurrent due to ambient illumination and i_{Amp} is the input referred noise of the amplifier. Hence, the overall signal to noise ratio is given by:

$$\frac{S}{N} = \frac{(RP_O)^2}{\left(2q(RP_O + i_{Amb})\Delta f + i_{Amp}^2 \Delta f\right)} \tag{5}$$

Leading to:

$$\frac{E_b R_b}{N_0} = \frac{(RP_O)^2}{\left(2q(RP_O + i_{Amb}) + i_{Amp}^2\right)} \tag{6}$$

This expression allows the bit-rate available R_b to be related to the range for a given required E_b/N_0. Then the modulation and detection scheme employed will determine the value of E_b/N_0 needed for a particular biterrorrate (BER). The value of the ambient light current varies considerably with the FOV and optical filtering conditions prevalent in the system. Therefore, for a 60-degree FOV, we take the pessimistic value of 1000 μA from [65], which assumes coarse optical filtering. The ambient light collected depends on $\sin^2(\theta_h)$ [69], and so we scale i_{Amb} appropriately as θ_h varies. Amplifier current noise is another quantity that varies somewhat depending on the device used in the optical RX. A typical device such as the Texas instruments OPA847 [69] offers a value of 2.7 pA Hz$^{-1/2}$, so we adopt a slightly more conservative value of 3 pA Hz$^{-1/2}$ in our calculations.

4.1.1.2. Millimeter-wave communications

We assume that the system will utilize 60 GHz as its frequency of transmission given the established products for this choice. Determination of the path loss for a 60 GHz mm-wave system is not straightforward, so for the outdoor scenario we use the International Telecommunication Union (ITU) LOS model [70]:

$$PL_{dB}(d) = 92.44 + 20\log_{10}(f) + 10n\log_{10}(d) \tag{7}$$

where d is the transmission distance in kilometers, f is the frequency in GHz and n is the path loss exponent, which is approximately two for this scenario. This is used in combination with the standard Friis model to produce a value of the received power as follows. In contrast with lower frequencies, the antennas will not be isotropic, and have numerical gains of G_T and G_R for TX and RX, respectively, so the overall received power in watts so for a transmitted signal power P_T will be:

$$P_R(d) = P_T G_T G_R 10^{-PL_{dB}(d)/10} \tag{8}$$

The antenna gains above are assumed to be equal, which provides asymmetric transmission system. Their values are determined based on the acceptance angle defined for optical transmission to give a fair comparison. For an ideal antenna, the gain is equal to the directivity, and so we can say that [71]:

$$G_T = G_R = \frac{4\pi}{\Omega_A} \tag{9}$$

where Ω_A is the beam area, taken to be the solid angle formed by the angle θ_h. As a result, we can write:

$$G_T = G_R = \frac{4\pi}{2\pi(1 - \cos\theta_h)} = \frac{2}{(1 - \cos\theta_h)} \tag{10}$$

We assume that the RX antenna feeds a matched preamplifier with noise factor F so that the signal to noise ratio S/N at the RX is:

$$\frac{S}{N} = \frac{P_R(d)}{FKTB} \tag{11}$$

where K is Boltzmann's constant and T is the temperature in Kelvin. For a bit-rate R_b average energy per bit E_b and noise power density N_0, we can then write:

$$\frac{E_b R_b}{N_0} = \frac{P_R(d)}{FKT} \tag{12}$$

4.1.1.3. Results

We assume the same transmission power for the links of 1 W and a light collection area of 15 cm diameter using an optical concentrator. This would have a value of θ_h equal to approximately 20 degrees. The noise factor is taken to be 5 dB as per the IEEE802.11ad standard giving $F = 3.2$. The results of the calculation are shown in **Figure 8** where it may be observed that FSO performs well for short distances, but millimeter waves offer better performance once the transmission distance exceeds 20–30 m. It may also be observed that for very short distances there is no appreciable free space loss (FSO) given the large collection area but no gain mechanism resulting in the flat portion of the characteristic until the loss begins to increase. We must also state that both systems could be impacted by adverse transmission conditions, rain for millimeter-wave and fog for FSO. Given the variability of these factors, the figure is intended to present a best-case comparison of the two systems.

4.1.2. Hotspot

The application scenario is a "hotspot" in a 3-meter high room as shown in **Figure 9**. As is apparent from the figure, the TX launches power within an emission cone with a half angle θ_h, and the RX has an acceptance angle that is also θ_h. This pairing of angles is optimum since a larger RX acceptance angle would mean that since terminals are transceivers, the uplink would transmit radiation that would miss the ceiling base station and vice versa. The geometry of this scenario means that for a hotspot radius d and range r:

$$\theta_h = \tan^{-1}(d/3); \quad r = \sqrt{d^2 + 9} \tag{13}$$

4.1.2.1. Optical wireless

The TX power is taken to be 10 W obtained from an LED lighting source, and a smaller RX diameter of 15 mm is used to represent the size possible on portable computing devices.

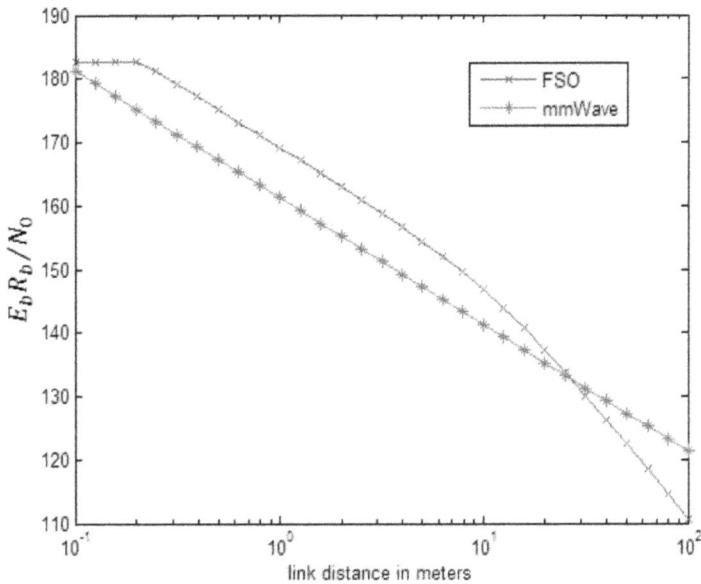

Figure 8. Comparison of FSO and millimeter-wave performance over a point-to-point link.

4.1.2.2. Millimeter-wave communications

In this example, we adopt the empirical model from the IEEE 802.11ad standard for a simple indoor LOS link [72]:

$$PL_{dB}(d) = 32.5 + 20\log_{10}(f) + 10n\log_{10}(d) \tag{14}$$

where d is the transmission distance in meters, f is the frequency in GHz and n is the path loss exponent, which is approximately two for this scenario. The antenna gains will again be given

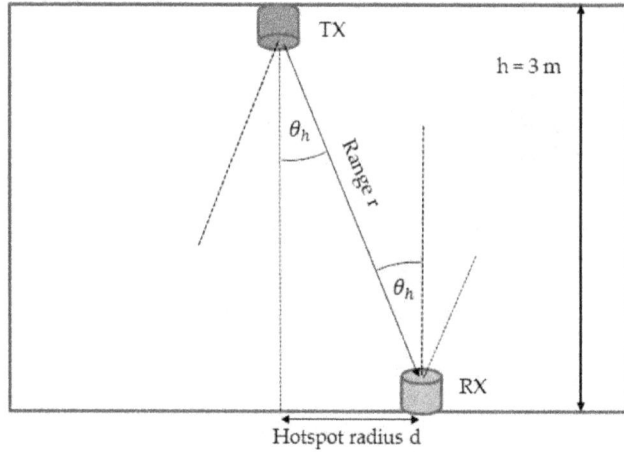

Figure 9. Geometry of the worst case hotspot alignment.

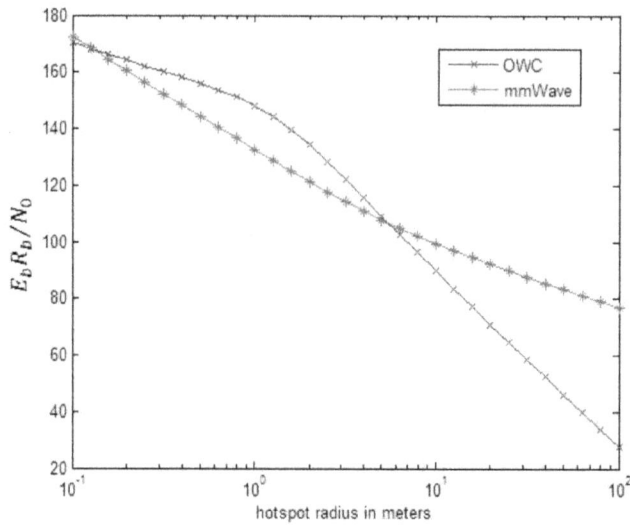

Figure 10. Comparison of OWC and 60 GHz performance for indoor hotspots.

by Eq. (10), but the angle will be that used for OWC obtained from Eq. (12). Here, the TX power is taken to be 10 mW since this represents a value that is permitted by many international standards [73] when one takes into consideration the antenna gains.

4.1.2.3. Results

In this indoor application, there will be no atmospheric losses, so the predictions will most likely be closer to the performance that could be seen using real links. The results obtained are shown in **Figure 10** and differ somewhat from the point-to-point case. Here, the change in FOV for both systems with hotspot radius assists their performance somewhat, particularly for the OWC link. It can be seen from the figure that there is a radius of up to a few meters where OWC is extremely competitive and could outperform millimeter-wave transmission. Furthermore, the infrastructure for the lighting will already be present so fewer extra components will be needed since the link uses the existing lights rather than a separate link.

5. Conclusions

This chapter has provided a brief introduction to potential 5G transmission systems based on optical and millimeter waves. It may be seen that both technologies have long histories and have been employed in military scenarios. Moreover, technological advances such as LiFi and increasing CMOS integration have brought both options to the fore in recent years. With respect to the future 5G integration, LOS transmission appears to be likely application given the significantly reduced performance of both methods once NLOS links are considered. Millimeter-wave transmission is probably more likely outdoors in most scenarios since its performance is increasingly superior to FSO once distances exceed a few tens of meters. It is the indoor sphere where OWC offers its best prospects for 5G because the LED lighting is becoming widespread. Thus, a relatively high-power visible light source is available for transmission as part of an office infrastructure. Transmission power at 60 GHz is restricted by international standard, especially when high antenna gains are employed as would be the case in small hotspots. Thus, OWC provides superior transmission performance over hotspots with diameters of a few meters, which is a realistic size. We have taken a simplified view of the application scenarios to compare them, and we acknowledge that significant developments are occurring, such as multiple input multiple output (MIMO) systems [74, 75]. Thus, the next stage of the future investigation is to incorporate these into the comparative modeling work. This can be coupled with experimental trials to determine the utility of OWC and millimeter waves in future 5G implementations.

Acknowledgements

This work was supported in part by Innovate UK under the grant ref. 103287.

Conflict of interest

The authors declare that they have no conflicts of interest.

Author details

Mark Stephen Leeson[1]* and Matthew David Higgins[2]

*Address all correspondence to: mark.leeson@warwick.ac.uk

1 School of Engineering, University of Warwick, Coventry, UK

2 WMG, University of Warwick, Coventry, UK

References

[1] European Project ICT-317669-METIS, Deliverable D8.4 METIS final Project Report Internet]. April 2015. Available from: https://www.metis2020.com/documents/deliverables/METIS_D8.4_v1.pdf [Accessed: 2018-02-20]

[2] Rappaport TS, Sun S, Mayzus R, Zhao H, Azar Y, Wang K, Wong GN, Schulz JK, Samimi M, Gutierrez F. Millimeter wave mobile communications for 5G cellular: It will work! IEEE Access. 2013;**1**:335-349. DOI: 10.1109/ACCESS.2013.2260813

[3] Ghassemlooy Z, Arnon S, Uysal M, Xu Z, Cheng J. Emerging optical wireless communications-advances and challenges. IEEE Journal on Selected Areas in Communications. 2015;**33**:1738-1749. DOI: 10.1109/JSAC.2015.2458511

[4] Borah DK, Boucouvalas AC, Davis CC, Hranilovic S, Yiannopoulos K. A review of communication-oriented optical wireless systems. EURASIP Journal on Wireless Communications and Networking. 2012;**2012**:91 28 p. DOI: 0.1186/1687-1499-2012-91

[5] Song J, Castellanos MR, Love DJ. Millimeter wave communications for 5G networks. In: Wong VWS, Schober R, Ng DWK, Wang L-C, editors. Key Technologies for 5G Wireless Systems. 1st ed. Cambridge: Cambridge University Press; 2017. pp. 188-213. DOI: 10.1017/9781316771655.011

[6] Elgala H, Mesleh R, Haas H. Indoor optical wireless communication: Potential and state-of-the-art. IEEE Communications Magazine. 2011;**49**:56-62. DOI: 10.1109/MCOM.2011.6011734

[7] Rangan S, Rappaport TS, Erkip E. Millimeter-wave cellular wireless networks: Potentials and challenges. Proceedings of the IEEE. 2014;**102**:366-385. DOI: 10.1109/JPROC.2014.2299397

[8] British Standards Institute. Safety of laser products, equipment classification, requirements and user's guide. BS EN 60825-1:2007. London: BSI; 2014

[9] Son IK, Mao S. A survey of free space optical networks. Digital Communications and Networks. 2017;**3**:67-77. DOI: 10.1016/j.dcan.2016.11.002

[10] O'Brien DC, Parry G, Stavrinou P. Optical hotspots speed up wireless communication. Nature Photonics. 2007;**1**:245-227. DOI: 10.1038/nphoton.2007.52

[11] Wisely DR, Neild I. A 100 Mb/s tracked optical telepoint. In: Proceedings of Personal, Indoor & Mobile Radio Communications (PIMRC 1997); Sept 1–4 1997; Helsinki. New York: IEEE; 1997. pp. 964-968

[12] Hranilovic S. Wireless Optical Communication Systems. 1st ed. New York: Springer; 2005. p. 197. DOI: 10.1007/b99592

[13] Higgins MD, Green RJ, Leeson MS. A genetic algorithm method for optical wireless channel control. Journal of Lightwave Technology. 2009;**27**:760-772. DOI: 10.1109/JLT.2008.928395

[14] Shiu DS, Kahn JM. Differential pulse-position modulation for power-efficient optical communication. IEEE Transactions on Communications. 1999;**47**:1201-1210. DOI: 10.1109/26.780456

[15] Chan VW. Optical satellite networks. Journal of Lightwave Technology. 2003;**21**:2811-2827. DOI: 10.1109/JLT.2003.819534

[16] Ghassemlooy Z, Popoola WO. Terrestrial free-space optical communications. In: Fares SA, Adachi F, editors. Mobile and Wireless Communications Network Layer and Circuit Level Design. 1st ed. London: InTech; 2010. pp. 356-392. DOI: 10.5772/7698

[17] Johnson LJ, Jasman F, Green RJ, Leeson MS. Recent advances in underwater optical wireless communication. Underwater Technology. 2014;**32**:167-175. DOI: 10.3723/ut.32.167

[18] Agrawal GP. Fiber-Optic Communication Systems. 3rd ed. New York: Wiley; 2009

[19] Bell AG. On the production and reproduction of sound by light. Journal of the Society of Telegraph Engineers. 1880;**9**:404-426. DOI: 10.1049/jste-1.1880.0046

[20] Goodwin E. A review of operational laser communication systems. Proceedings of the IEEE. 1970;**58**:1746-1752. DOI: 10.1109/PROC.1970.7998

[21] Begley DL. Free-space laser communications: A historical perspective. In: Proceedings of the 15th Annual Meeting of the IEEE Lasers and Electro-Optics Society (LEOS); 10–14 Nov. 2002; Glasgow. New York: IEEE; 2003. DOI: 10.1109/LEOS.2002.1159343

[22] Leitgeb E. Future applications of optical wireless and combination scenarios with RF technology. In: Proceedings of MIPRO; May 22–26 2017. Opatija; Kruzna: MIPRO; 2017. pp. 513-516

[23] Libich J, Komanec M, Zvanovec S, Pesek P, Popoola WO, Ghassemlooy Z. Experimental verification of an all-optical dual-hop 10 Gbit/s free-space optics link under turbulence regimes. Optics Letters. 2015;**40**:391-394. DOI: 10.1364/OL.40.000391

[24] Gfeller FR, Bapst U. Wireless in-house data communication via diffuse infrared radiation. Proceedings of the IEEE. 1979;**67**:1474-1486. DOI: 10.1109/PROC.1979.11508

[25] McCullagh MJ, Wisely DR. 155 mbit/s optical wireless link using a bootstrapped silicon APD receiver. Electronics Letters. 1994;**30**:430-432. DOI: 10.1049/el:19940308

[26] Carruthers J, Kahn JM. Angle diversity for nondirected wireless infrared communication. IEEE Transactions on Communications. 2000;**48**:960-969. DOI: 10.1109/26.848557

[27] Green RJ, Joshi H, Higgins H, Leeson MS. Recent developments in indoor optical wireless systems. IET Communications. 2008;**2**:3-10. DOI: 10.1049/iet-com:20060475

[28] Nirmalathas A, Wang K, Lim C, Wong E, Skafidas E, Alumeh K, Song T. Liang T multi-gigabit indoor optical wireless networks — Feasibility and challenges. In: Proceedings of IEEE Photonics Society Summer Topical Meeting Series; 11–13 July 2016; Newport Beach. New York: IEEE; 2016. pp. 130-131

[29] Cho J, Park JH, Kim JK, Schubert EF. White light-emitting diodes: History, progress, and future. Laser & Photonics Reviews. 2017;**11**:1600147. DOI: 10.1002/lpor.201600147

[30] Komine T, Nakagawa M. Fundamental analysis for visible light communication system using LED lights. IEEE Transactions on Consumer Electronics. 2004;**50**:100-107. DOI: 10.1109/TCE.2004.1277847

[31] Tae-Gyu K. Visible-light communications. In: Arnon S, Barry J, Karagiannidis G, Schober R, Uysal M, editors. Advanced Optical Wireless Communication Systems. 1st ed. Cambridge: Cambridge University Press; 2012. pp. 351-368

[32] Cossu G, Khalid AM, Choudhury P, Corsini R, Ciaramella E. 3.4 Gbit/s visible optical wireless transmission based on RGB LED. Optics Express. 2012;**20**:B501-B506. DOI: 10.1364/OE.20.00B501

[33] Dimitrov S, Haas H. Principles of LED Light Communications. 1st ed. Cambridge: Cambridge University Press; 2015 206 p

[34] Haas H, Chen C. Visible light communication in 5G. In: Wong VWS, Schober R, Ng DWK, Wang L-C, editors. Key Technologies for 5G Wireless Systems. 1st ed. Cambridge: Cambridge University Press; 2017. pp. 188-213. DOI: 10.1017/9781316771655.015

[35] Haas H, Yin L, Wang Y, Chen C. What is LiFi? Journal of Lightwave Technology. 2016;**34**: 1533-1544. DOI: 10.1109/JLT.2015.2510021

[36] Yuan R, Ma J. Review of ultraviolet non-line-of-sight communication. China Communications. 2016;**13**:63-75. DOI: 10.1109/CC.2016.7513203

[37] Shaw GA, Siegel AM, Model J. Extending the range and performance of non-line-of-sight ultraviolet communication links. In: Proc. SPIE 6231, Unattended Ground, Sea, and Air Sensor Technol. and Appl. VIII; 2 May 2006; Orlando. Bellingham: SPIE. p. 12

[38] Drost RJ, Sadler BM. Survey of ultraviolet non-line-of-sight communications. Semiconductor Science and Technology. 2014;29:084006. DOI: 10.1088/0268-1242/29/8/084006

[39] Niu Y, Li Y, Jin D, Su L, Vasilakos AV. A survey of millimeter wave communications (mmWave) for 5G: Opportunities and challenges. Wireless Networks. 2015;21:2657-2676. DOI: 10.1007/s11276-015-0942-z

[40] Rappaport TS, Murdock JN, Gutierrez F. State of the art in 60-GHz integrated circuits and systems for wireless communications. Proceedings of the IEEE. 2011;99:1390-1436. DOI: 10.1109/JPROC.2011.2143650

[41] Saha SK, Vira VV, Garg A, Koutsonikolas A. A feasibility study of 60 GHz indoor WLANs. In: Proceedings of 25th International Conference on Computer Communication and Networks (ICCCN); 1–4 August 2016; Waikoloa. New York: IEEE; 2016. pp. 1-9

[42] Panagopoulos AD, Arapoglou P-DM, Cottis PG. Satellite communications at KU, KA, and V bands: Propagation impairments and mitigation techniques. IEEE Communication Surveys and Tutorials. 2004;6:2-14. DOI: 10.1109/COMST.2004.5342290

[43] Kim J, Lee J-J, Lee W. Strategic control of 60 GHz millimeter-wave high-speed wireless links for distributed virtual reality platforms. Mobile Information Systems. 2017;2017: 5040347. DOI: 10.1155/2017/5040347

[44] Patole S, Torlak M, Wang D, Ali M. Automotive radars: A review of signal processing techniques. IEEE Signal Processing Magazine. 2017;34:22-35. DOI: 10.1109/MSP.2016.2628914

[45] Nanzer JA. Microwave and Millimeter-Wave Remote Sensing for Security Applications. 1st ed. Boston: Artech House; 2012

[46] Logani MK, Bhopale MK, Ziskin MC. Millimeter wave and drug induced modulation of the immune system-application in Cancer immunotherapy. Journal of Cell Science and Therapy. 2011;S5:002. DOI: 10.4172/2157-7013.S5-002

[47] Rappaport TS. Wireless Communications: Principles and Practice. 2nd ed. Prentice Hall: Upper Saddle River; 2002. p. 736

[48] Qiao J, Shen X, Mark JW, He Y. MAC-layer concurrent beamforming protocol for indoor millimeter-wave networks. IEEE Transactions on Vehicular Technology. 2015;64:327-338. DOI: 10.1109/TVT.2014.2320830

[49] Wiltse JC. History of millimeter and submillimeter waves. IEEE Transactions on Microwave Theory and Techniques. 1984;MTT-32:1118-1127. DOI: 10.1109/TMTT.1984.1132823

[50] Daniels RC, Heath RW. 60 GHz wireless communications: Emerging requirements and design recommendations. IEEE Vehicular Technology Magazine. 2007;2:41-50. DOI: 10.1109/MVT.2008.915320

[51] Wu X, Wang C-X, Sun J, Huang J, Feng R, Yang Y, Ge X. 60-GHz millimeter-wave channel measurements and modeling for indoor office environments. IEEE Transactions on Antennas and Propagation. 2017;65:1912-1924. DOI: 10.1109/TAP.2017.2669721

[52] Bains AS. An overview of millimeter wave communications for military applications. Defence Science Journal. 1993;**43**:27-36. DOI: 10.14429/dsj.43.4492

[53] Pi Z, Khan F. An introduction to millimeter-wave mobile broadband systems. IEEE Communications Magazine. 2011;**49**:101-107. DOI: 10.1109/MCOM.2011.5783993

[54] Ninomiya T, Saito T, Ohashi Y, Yatsuka H. 60-GHz transceiver for high-speed wireless LAN system. In: Proceedings of IEEE MTT-S Int. Microw. Symp, 17–21 June 1996; San Francisco. New York: IEEE; 2002. pp. 1171-1174

[55] Doan CH, Emami S, Niknejad AM, Brodersen RW. Design of CMOS for 60 GHz applications. In: Proceedings of IEEE the Int. Solid-State Circuits Conf.; 15–19 February 2004; San Francisco. New York: IEEE; 2004. pp. 440-538

[56] Alldred D, Cousins B, Voinigescu SP. A 1.2 V, 60-GHz radio receiver with on-chip transformers and inductors in 90-nm CMOS. In: Proceedings of the Compound Semiconductor Integrated Circuit Symposium (CSIC 2006); 12–15 November 2006; San Antonio. New York: IEEE; 2006. pp. 51-54. DOI: 10.1109/CSICS.2006.319876

[57] Dawn D, Sen P, Sarkar S, Perumana B, Pinel S, Laskar J. 60-GHz integrated transmitter development in 90-nm CMOS. IEEE Transactions on Microwave Theory and Techniques. 2009;**57**:2354-2367. DOI: 10.1109/TMTT.2009.2029028

[58] Tomkins A, Aroca RA, Yamamoto T, Nicolson ST, Doi Y, Voinigescu SP. A zero-IF 60 GHz 65 nm CMOS transceiver with direct BPSK modulation demonstrating up to 6 Gb/s data rates over a 2 m wireless link. IEEE Journal of Solid-State Circuits. 2009;**44**:2085-2099. DOI: 10.1109/JSSC.2009. 2022918

[59] Wu R, Minami R, Tsukui Y, Kawai S, Seo Y, Sato S, Kimura K, Kondo S, Ueno T, Fajri N, Maki S, Nagashima N, Takeuchi Y, Yamaguchi T, Musa A, Tokgoz KK, Siriburanon T, Liu B, Wang Y, Pang J, Li N, Miyahara M, Okada K, Matsuzawa A. 64-QAM 60-GHz CMOS transceivers for IEEE 802.11ad/ay. IEEE Journal of Solid-State Circuits. 2017;**52**:2871-2891. DOI: 10.1109/JSSC.2017.2740264

[60] Widrow B. Adaptive antenna systems. Proceedings of the IEEE. 1967;**55**:2143-2159. DOI: 10.1109/PROC.1967.6092

[61] Mietzner J, Schober R, Lampe L, Gerstacker WH, Hoeher PA. Multiple-antenna techniques for wireless communications–a comprehensive literature survey. IEEE Communication Surveys and Tutorials. 2009;**11**:87-105. DOI: 10.1109/SURV.2009.090207

[62] Kutty S, Sen D. Beamforming for millimeter wave communications: An inclusive survey. IEEE Communication Surveys and Tutorials. 2016;**18**:949-973. DOI: 10.1109/COMST.2015. 2504600

[63] Van Veen BD, Buckley KM. Beamforming: A versatile approach to spatial filtering. IEEE ASSP Magazine. 1988;**5**:4-24. DOI: 10.1109/53.665

[64] Fettweis G, Zimmermann E, Allen B, O'Brien D C, Chevillat P editors. Short-range wireless communications. In: Tafazolli, Editor. Technologies for the Wireless Future. 1st ed. Chichester: Wiley; 2006. p. 227-312. DOI: /10.1002/0470030453.ch7

[65] Wolf M, Kress D. Short-range wireless infrared transmission: The link budget compared to RF. IEEE Wireless Communications. 2003;**10**:8-14. DOI: 10.1109/MWC.2003.1196397

[66] Xiao M, Mumtaz S, Huang Y, Dai L, Li Y, Matthaiou M, Karagiannidis GK, Björnson E, Yang K, Chih-Lin I, Ghosh A. Millimeter wave communications for future mobile networks. IEEE Journal on Selected Areas in Communications. 2017;**5**:1909-1935. DOI: 10.1109/JSAC.2017.2719924

[67] Kahn KM, Barry JR. Wireless infrared communications. Proceedings of the IEEE. 1997;**5**: 265-298. DOI: 10.1109/5.554222

[68] Pakravan MR, Simova E, Kavehrad M. Holographic diffusers for indoor infrared communication systems. International Journal of Wireless Information Networks. 1997;**4**:259-274. DOI: 10.1023/A:1018876326494

[69] Otte R, de Jong LP, van Roermund AHM. Low-Power Wireless Infrared Communications. 1st ed. Boston: Kluwer; 1999. p. 159. DOI: 10.1007/978-1-4757-3015-9_3Texas Instruments. Wideband Operational Amplifier [Internet]. 2008. Available from :http://www.ti.com/lit/ds/symlink/opa847.pdf [Accessed 2018-02-23]

[70] ITU Recommendation ITU-R P.2001–2 [Internet]. 2015. Available from https://www.itu.int/dms_pubrec/itu-r/rec/p/R-REC-P.2001-2-201507-I!!PDF-E.pdf [Accessed 2018–02-23]

[71] Kraus JD, Marhefka RJ. Antennas for all Applications. 3rd ed. New York: Wiley; 2001 960 p

[72] Maltsev E. Perahia R, Maslennikov A, Lomayev A, Khoryaev A, Sevastyanov A. Path loss model development for TGad channel models, IEEE 802.11–09/0553r1 [Internet]. 2009. Available from: https://mentor.ieee.org/802.11/dcn/09/11-09-0553-01-00ad-path-loss-model-development-for-tgad-channel-models.ppt [Accessed 2018–02-23]

[73] Smulders P. Exploiting the 60 GHz band for local wireless multimedia access: Prospects and future directions. IEEE Communications Magazine. 2002;**40**:140-147. DOI: 10.1109/35.978061

[74] Lian J, Brandt-Pearce M. Multiuser MIMO indoor visible light communication system using spatial multiplexing. Journal of Lightwave Technology. 2017;**23**:5024-5033, Dec.1, 1 2017. DOI: 10.1109/JLT.2017.2765462

[75] Sun S, Rappaport TS, Heath RW, Nix A, Rangan S. Mimo for millimeter-wave wireless communications: Beamforming, spatial multiplexing, or both? IEEE Communications Magazine. 2014;**52**:110-121. DOI: 10.1109/MCOM.2014.6979962

Chapter 3

Rateless Space-Time Block Codes for 5G Wireless Communication Systems

Ali Alqahtani

Additional information is available at the end of the chapter

http://dx.doi.org/10.5772/intechopen.74561

Abstract

This chapter presents a rateless space-time block code (RSTBC) for massive multiple-input multiple-output (MIMO) wireless communication systems. We discuss the principles of rateless coding compared to the fixed-rate channel codes. A literature review of rateless codes (RCs) is also addressed. Furthermore, the chapter illustrates the basis of RSTBC deployments in massive MIMO transmissions over lossy wireless channels. In such channels, data may be lost or are not decodable at the receiver end due to a variety of factors such as channel losses or pilot contamination. Massive MIMO is a breakthrough wireless transmission technique proposed for future wireless standards due to its spectrum and energy efficiencies. We show that RSTBC guarantees the reliability of the system in such highly lossy channels. Moreover, pilot contamination (PC) constitutes a particularly significant impairment in reciprocity-based multi-cell systems. PC results from the non-orthogonality of the pilot sequences in different cells. In this chapter, RSTBC is also employed in the downlink transmission of a multi-cell massive MIMO system to mitigate the effects of signal-to-interference-and-noise ratio (SINR) degradation resulting from PC. We conclude that RSTBC can effectively mitigate such interference. Hence, RSTBC is a strong candidate for the upcoming 5G wireless communication systems.

Keywords: massive MIMO, rateless codes, STBC, pilot contamination, 5G

1. Introduction

In practice, the transmitted data over the channel are usually affected by noise, interference, and fading. Several channel models, such as additive white Gaussian noise (AWGN), binary symmetrical channel (BSC), binary erasure channel (BECH), wireless fading channel, and lossy (or erasure) channel, are introduced in which errors' (or losses) control technique is required to reduce the errors (or losses) caused by such channel impairments [1].

This technique is called *channel coding*, which is a main part of the digital communication theory. Historical perspective on channel coding is given in [2]. Generally speaking, channel coding, characterized by a code rate, is designed by controlled-adding redundancy to the data to detect and/or correct errors and, hence, achieve reliable delivery of digital data over unreliable communication channels. Error correction may generally be realized in two different error control techniques, namely: forward error correction (FEC), and backward error correction (BEC). The former omits the need for the data retransmission process, while the latter is widely known as automatic repeat request (or sometimes automatic retransmission query) (ARQ).

For large size of data, a large number of errors will occur, and thereby, it is difficult for FEC to work reasonably. The ARQ technique, in such conditions, requires more retransmission processes, which will cause significant growth in power consumption. However, these processes will sustain additional overhead that includes data retransmission and adding redundancy into the original data. They cannot correctly decode the source data when the packet loss rate is high [3]. Therefore, it is of significant interest to design a simple channel coding with a flexible code rate and capacity approaching behavior to achieve robust and reliable transmission over universal lossy channels. Rateless codes constitute such a class of schemes. We describe the concept of rateless coding in the next section.

2. Concept of rateless codes

Rateless codes (RCs) are channel codes used for encoding data to generate incremental redundancy codes and then, are transmitted over channels with variable packet error rate. The interpretation of the terminology "rateless" is that the code does not fix its code rate before transmission. Rather, it can only be determined after correctly recovering the transmitted data. In the available literature, the rateless code is typically referred to by some associated terminologies such as "variable-rate," "rate-compatible," "adaptive-rate," or "incremental redundancy" scheme [4]. However, the rate of a rateless code can be considered in two perspectives as the instantaneous rate and the effective rate. The instantaneous rate is the ratio of the number of information bits to the total number of bits transmitted at a specific instant. On the other hand, the effective rate is the rate realized at the specific point when the codeword has been successfully received [5].

The counterpart of rateless coding is fixed-rate coding, which is basically well known in the literature of channel codes. The relationship between rateless and fixed-rate channel codes can be seen as the correspondence between continuous and discrete signals or the construction of a video clip from video frames. In this illustrating analogy, the fixed-rate code corresponds to the discrete signal or to the video frame, while the rateless code is viewed as the continuous signal or the video clip [5]. Basically, rateless codes are proposed to solve the problem of data packet losses. They can continuously generate potentially unlimited number of data streams until an acknowledgment from the receiver is received declaring successful decoding.

The basic concept of rateless codes is illustrated in **Figure 1** [1]. From the figure, a total of k_c packets, obtained from the fragmented source data, are encoded by the transmitter to get a

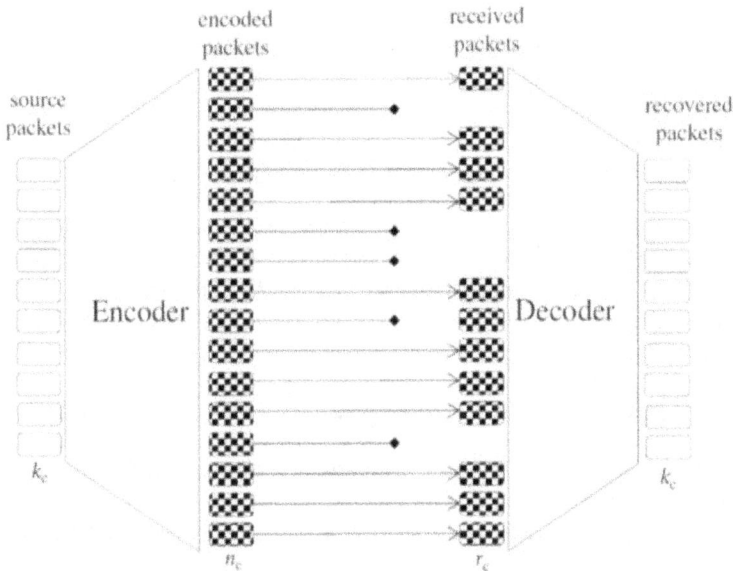

Figure 1. General encoding and decoding processes of rateless codes [1].

large number of encoded packets n_c. Due to the lossy channel, several encoded packets are lost during the transmission, and finally, only r_c encoded packets are collected by the receiver. The decoding process on the received packets should be able to recover all original k_c packets.

To illustrate the importance of rateless codes, let's assume that we have a fixed-rate code C_{fixed} of fixed-code rate R_{fixed} designed to achieve a performance close to the channel capacity target C_{target} at a specific signal-to-noise-ratio (SNR), φ_{fixed}. However, the channel fluctuations make the fixed-rate code impose two limitations [1]. First, if the actual SNR at the receiver is actually greater than φ_{fixed}, then the code essentially becomes an inefficient channel code. That is because the code incorporates more redundancy than the actual channel conditions require. Second, on the other hand, if the actual SNR becomes lower than φ_{fixed}, then the channel will be in an outage for the reason that the fixed-rate code C_{fixed} no longer provides sufficient redundancy appropriate for the actual channel conditions. Contrasting with fixed-rate code, the rateless code has a flexible code rate in accordance with the channel quality, which is time varying in nature. Another benefit of RCs is that they potentially do not require the channel state information (CSI) at the transmitter. This property is of particular importance in the design of codes for wireless channels. In particular, RCs can be employed in multi-cell cellular systems when channel estimation errors severely degrade the performance.

3. Rateless coding and hybrid automatic retransmission query

For further discussion, there is an analogy between hybrid ARQ (HARQ) and rateless codes, since they transmit additional symbols until the received information is successfully decoded. On the other hand, they do have some differences. HARQ refers to a special transmission mechanism, which combines the conventional ARQ principle with error control coding. The basic three ARQ protocols are stop-and-wait ARQ, go-back-N ARQ, and selective repeat ARQ. All the three ARQ protocols use the sliding window protocol to inform the transmitter on which data frames or symbols should be retransmitted. **Figure 2** illustrates the ARQ schemes: stop-and-wait ARQ (half duplex), continuous ARQ with pullback (full duplex), and continuous ARQ with selective repeat (full duplex). In each of them, time is advancing from lift to right [6].

These protocols reside in the data link or transport layers of the open systems interconnection (OSI) model. This is one difference between the proposed RSTBC and HARQ, since RSTBC is employed in the physical layer. Comparing rateless codes to HARQ, we summarize the following points:

1. RC is often viewed as a form of continuous incremental redundancy HARQ in the literature [5].

2. HARQ is not capable of working over the entire SNR range, and therefore, it necessitates combination with some form of adaptive modulation and coding (AMC). On the other hand, RC can entirely eliminate AMC and work over a wide range of SNR [7].

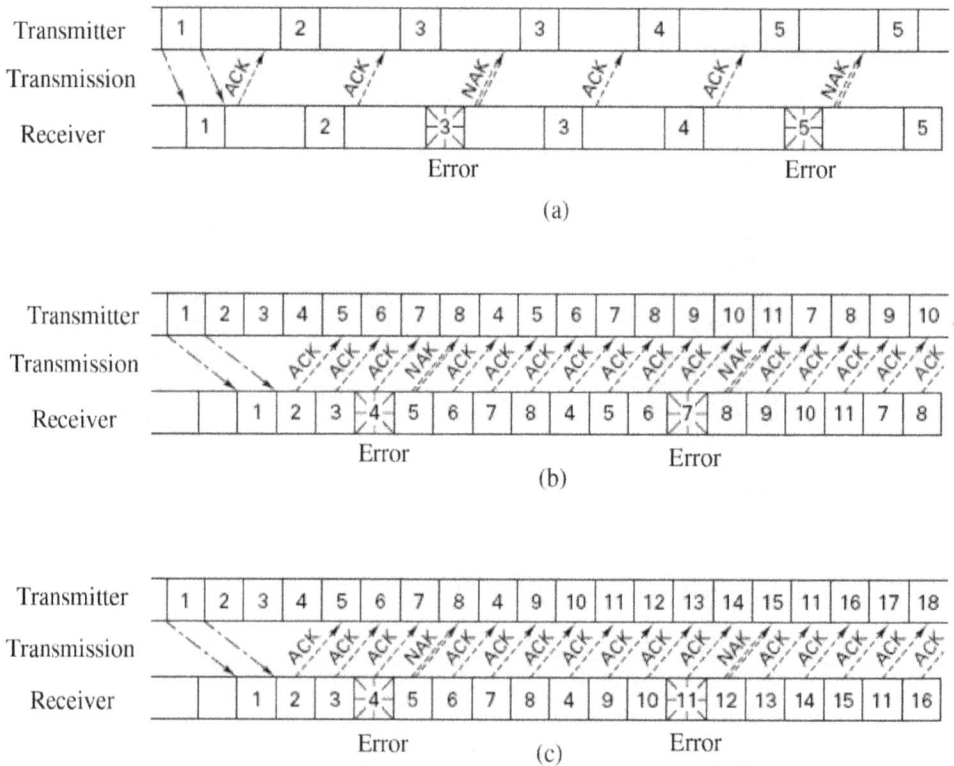

Figure 2. Automatic repeat request (ARQ) [6]. (a) Stop and wait ARQ (half duplex); (b) continuous ARQ with pullback (full duplex); (c) continuous ARQ with selective repeat (full duplex).

3. From the point of view of redundancy, HARQ has more redundancy, since it requires many acknowledgments (ACK) or negative acknowledgments (NACK) for each packet transmission return to show successful/unsuccessful decoding, respectively. In contrast, only a single-bit acknowledgment is needed for the transmission of a message with RC [8]. When the number of receivers is large, ARQ acknowledgments may cause significant delays and bandwidth consumption. Consequently, using ARQ for wireless broadcast is not scalable [9].

4. It was seen in [8] that RC is capable of outperforming ARQ completely at low SNRs in broadcast communication. However, they behave the same in point to point as well as in high-SNR broadcast communications.

5. RC and the basic ARQ differ in code construction. RCs can generate different redundant blocks, while ARQ merely retransmits the same block [8]. For different receivers, distinct and independent errors are often encountered. In such cases, the merely retransmitted data packets are only useful to a specific user while they are with no value for others. Hence, it is highly undesirable to send respective erroneous data frames or symbols to each user.

6. The physical layer RCs are useful since the decoder can exploit useful information from packets that are dropped by ARQ protocols in higher layers [7].

4. Rateless codes' literature review

In the past decade, rateless codes have gained a lot of concerns in both communication and information theory research communities, which led to the strong theory behind these codes mostly for erasure channels. Most of the available works in the rateless codes literature are extensions of the fountain codes over the erasure channels [10]. The name "fountain" came from the analogy to a water supply capable of giving an unlimited number of water drops. Due to this reason, rateless codes are also referred to as fountain codes. They were initially developed to achieve efficient transmission in erasure channels, to which the initial work on rateless codes has mainly been limited, with the primary application in multimedia video streaming [10].

The first practical class of rateless codes is the Luby Transform (LT) code which was originally intended to be used for recovering packets that are lost (erased) during transmission over computer networks. The fundamentals of LT are introduced in [11] in which the code is rateless since the number of encoded packets that can be generated from the original packets is potentially limitless. **Figure 3** illustrates the block diagram of LT encoder. The original packets can be generated from slightly larger encoded packets. Although the encoding process of LT is quite simple, however, LT requires the proper design of the degree distribution (Soliton distribution-based) which significantly affects the performance of the code.

Afterward, LT code was extended to the well-known Raptor code [12] by appending a weak LT encoder with an outer pre-code such as the irregular low-density parity check code (LDPC). **Figure 4** depicts the general block diagram of Raptor code.

The decoding algorithm of the Raptor code depends on the decoder of the LT code and the pre-code used. However, the Raptor code requires lesser overhead. But it has disadvantages such as the lower bound of the total overhead depends on the outer code and the decoding algorithm implementation is slightly more complicated due to multiple decoding processes.

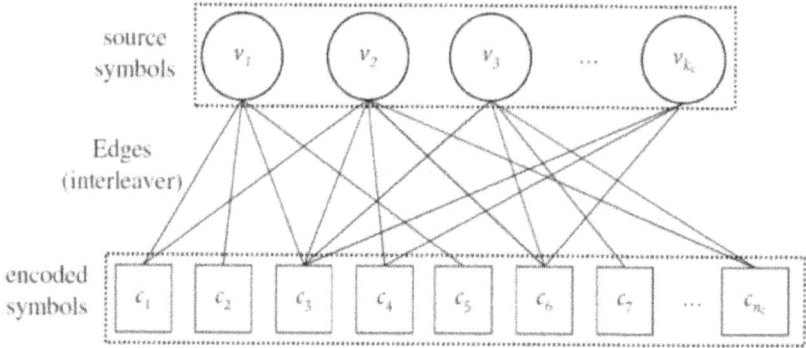

Figure 3. Block diagram of the LT encoder [5].

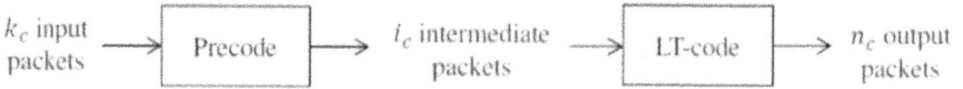

Figure 4. Block diagram of the Raptor code encoder [5].

Online codes [13] also belong to the family of fountain rateless codes and work based on two layers of packet processing (inner and outer). However, in contrast to the LT and Raptor codes, online codes have more encoding and decoding complexity as a function of the block length. The overall design of the online code is shown in **Figure 5**. LT and Raptor codes were originally intended to be used for transmission over the BEC channel such as Internet channel, where the transmitted packets are erased by routers along the path.

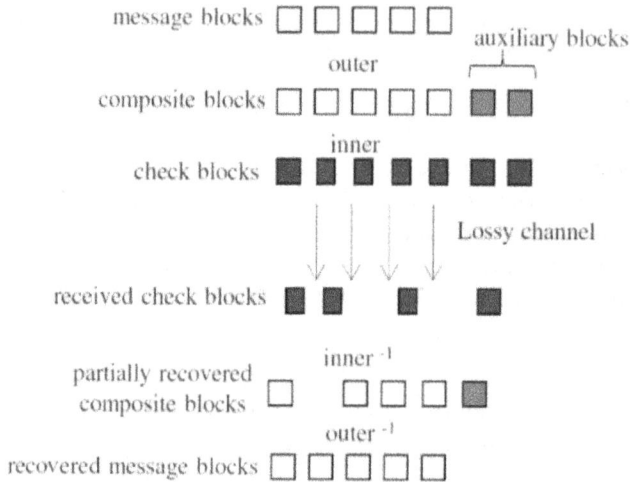

Figure 5. Online code encoding and decoding design [13].

On the other hand, some works have studied their performance on noisy channels such as BSC and AWGN channels [14]. Although it was demonstrated that the Raptor codes have better performance on a wide variety of noisy channels than LT codes, however, both schemes exhibit high error floors in such channels. The previous rateless codes have fixed-degree distribution, which causes degradation in performance when employing over noisy channels.

Motivated by this result, a reconfigurable rateless code was introduced in [5] which adaptively can modify its encoding/decoding algorithms by adjusting its degree distribution incrementally according to the channel conditions. Such code is not systematic and remains fixed if no new knowledge of channel condition is obtained from feedback. By dropping this assumption, as in [15], the significant overhead reduction can be achieved while still maintaining the same encoding and decoding complexities. In addition, the effective code rate of [5] is actually determined by the decoder, not the encoder.

In another perspective, the use of rateless codes in the physical layer can be beneficial since the decoder can exploit useful information even from packets that cannot be correctly decoded and therefore are ignored by higher layers [7]. A construction of physical-layer Raptor codes based on protographs was proposed in [16].

Other works of rateless coding over the AWGN channel were provided in [17, 18]. For wireless channels, rateless code paradigm was found in many works. In [19], a rateless coding scheme based on traditional Raptor code is introduced in a single-input single-output (SISO) system over fading channels. A similar approach is presented in [20] by the same authors for relay networks. The authors in [21] have considered one of the latest works of rateless coding over wireless channels. They tackle the high error floor problem arising from the low-density generator matrix (LDGM)-like encoding scheme in conventional rateless codes.

While there are significant works on rateless codes for AWGN channels, few work exists on rateless codes for MIMO systems. Rateless codes for MIMO channels were introduced in [22], where two rateless code constructions were developed. The first one was based on simple layering over an AWGN channel. The second construction used a diagonal layering structure over a particular time-varying scalar channel. However, the latter is merely concatenating a rateless code (outer code) using dithered repetition with the vertical Bell Labs layered space-time (V-BLAST) code (inner code) [23].

Away from digital fountain codes, discussed so far, performance limits and code construction of block-wise rateless coding for conventional MIMO fading channels are studied in [24]. The authors have used the diversity multiplexing tradeoff (DMT) as a performance metric [25]. Also, they have demonstrated that the design principle of rateless codes follows the approximately universal codes [26] over MIMO channels. In addition, simple rateless codes that are DMT optimal for a SISO channel have also been examined. However, [24] considered the whole MIMO channel as parallel sub-channels, in which each sub-channel is a MIMO channel. Furthermore, for each block, the code construction of symbols within the redundant block is not discussed. Hence, more investigation of other performance metrics for the scheme proposed in [24] under different channel scenarios is required. In [27], a cognitive radio network employs rateless coding along with queuing theory to maximize the capacity of the secondary user while achieving primary users' delay requirement. Furthermore, [28] presents a novel framework of opportunistic beam-forming employing rateless code in multiuser MIMO downlink to provide faster and high quality of service (QoS) wireless services.

5. Rateless codes applications

There are various applications of rateless codes:

- Video streaming over the Internet and packet-based wireless networks: The application of rateless codes to video streaming was initially proposed for multimedia broadcast multicast system (MBMS) standard of the 3GPP [29, 30].

- Broadcasting has been extensively used in wireless networks to distribute information of universal attention, for example, safety warning messages, emergency information, and weather information, to a large number of users [31, 32]. Rateless coding has been utilized in the 3GPP's Release 6 multimedia broadcast/multicast service (MBMS) [33].

- Wearable wireless networks: A wearable body area network (WBAN) is an emerging technology that is developed for wearable monitoring application. Wireless sensor networks are usually considered one of the technological foundations of ambient intelligence. Agile, low-cost, ultra-low power networks of sensors can collect a massive amount of valuable information from the surrounding environment [34, 35]. Wireless sensor network (WSN) technologies are considered one of the key research areas in computer science and the health-care application industries for improving the quality of life. A block-based scheme of rateless channel erasure coding was proposed in [36] to reduce the impact of wireless channel errors on the augmented reality (AR) video streams, while also reducing energy consumption.

6. Motivation to rateless space-time coding

According to literature survey, there is not enough research work yet on rateless space-time codes (STCs), even for the regular MIMO systems. Few works in rateless STCs are available such as [37, 38]. In [37], a rateless coding scheme was introduced for the AWGN channel, using layering, repetition and random dithering. The authors also extended their work to multiple-input single-output (MISO) Gaussian channels where the MISO channel is converted to parallel AWGN channels. In [38], the performance of MIMO radio link is improved by a rate-varying STC under a high-mobility communication system. Rateless coding can be extended to space-time block codes (STBCs), where the coding process is performed blockwise over time and space. The main advantage of STBCs is that they can provide full diversity gain with relatively simple encoding and decoding schemes. Unlike the conventional fixed-rate STBC, rateless STBC is designed such that the code rate is not fixed before transmission. Instead, it depends on the instantaneous channel conditions.

Incorporating RSTBC in massive MIMO systems is reasonable and very attractive, since rateless coding is based on generating a massive number of encoded blocks, and massive MIMO technique uses a large number of antenna elements. Motivated by such fact, in this chapter, a new approach has been developed to fill the gap between rateless STCs and massive MIMO systems by exploiting significant degrees of freedom available in massive MIMO systems for rateless coding. The contribution of RSTBC is to convert lossy massive MIMO channels into lossless ones and provide a reliable robust transmission when very large MIMO dimensions are used.

7. Rateless space-time block code for massive MIMO systems

Massive MIMO wireless communication systems have been targeted for deployment in the fifth-generation (5G) cellular standards, to enhance the wireless capacity and communication reliability [39] fundamentally. In massive MIMO systems, a large number of antennas, possibly hundreds or even thousands, work together to deliver big data to the end users. Despite the significant enhancement in capacity and/or link quality offered by MIMO systems and space-time codes (STCs) [40, 41], it has been shown recently that massive MIMO can even improve the performance of MIMO systems dramatically. This has prompted a lot of research works on massive MIMO systems lately.

In this section, we illustrate the mechanism of rateless space-time block code (RSTBC) in a massive MIMO system, as we have addressed in [42–45]. **Figure 6** shows simply the encoding and decoding processes, where a part of the encoded packets (or blocks) cannot be received due to channel losses. Hence, with the availability of slightly larger encoded packets, the receiver can recover the original packets from the minimum possible number of transmitted encoded packets that are already received. The required number of blocks for recovery depends on the loss rate of the channel. During the transmission, the receiver of a specific user measures the mutual information between the transmitter and the receiver after it receives each block and compares it to a threshold.

Namely, it is desired to decode a message of total mutual information M. Assume that the required packets to deliver the message correctly are $[X_1\ X_2\ \cdots\ X_l\ \cdots\ X_L]$, where X_l is the $N_t \times T$ codeword matrix transmitted during the lth block, N_t is the number of transmit antenna elements at the base station (BS), T is the number of time slots, and L is the number of required blocks at the receiver to recover the transmitted block. Let m_l denote the measured mutual information after receiving the codeword block X_l. If $m_l \leq M$, the receiver waits for further blocks, else if $m_l > M$, the receiver sends a simple feedback to the transmitter to stop transmitting the remaining part of the encoded packets and remove them from the BS buffer. This process continues until the receiver accumulates enough number of blocks (L) to recover the message or the time allowed is over the channel coherence time. The decoding process is conducted sequentially, first using X_1, then $[X_1\ X_2]$ if X_1 is not sufficient, and so forth. Once the check-sum condition is satisfied, the received blocks are linearly combined at the receiver to decode the whole underling message. It should be noted that the code is described as "rateless" because the number of required blocks (L) to recover the message is not fixed before transmission; rather, it depends on the channel state. The dimensions in which the code is extended ratelessly are time (number of channel uses) and space (number of functional antennas) as well as it belongs to block codes. Therefore, it is called rateless space-time block code (RSTBC).

Before proceeding, each of the RSTBC matrix X_l is constructed based on the following random process

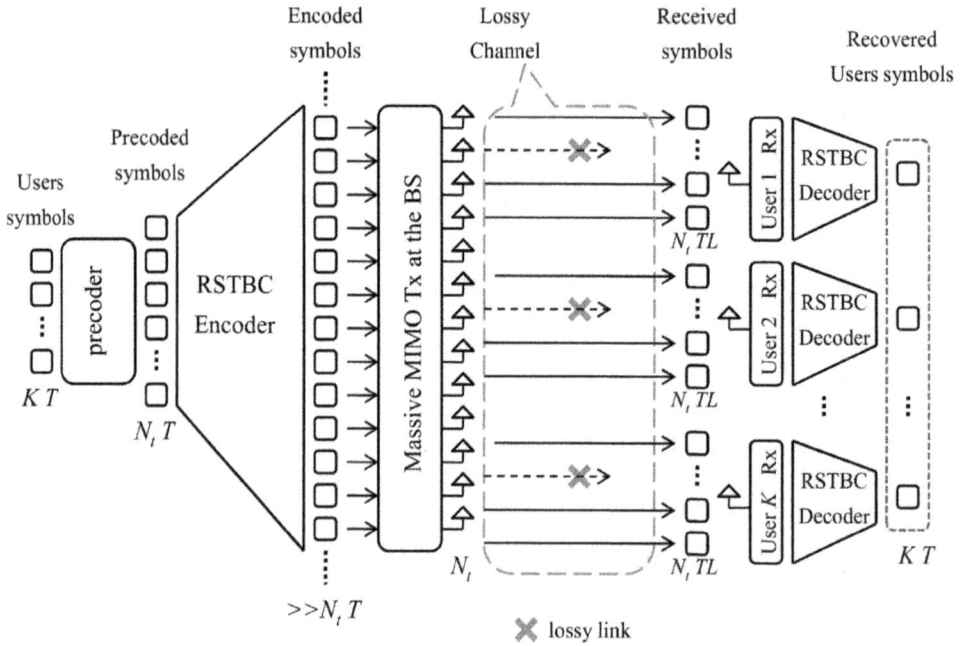

Figure 6. Encoding and decoding processes of RSTBC in a massive MIMO system.

$$X_l = X \odot D_l \tag{1}$$

where $\odot$ denotes the element-wise multiplication operation (Hadamard product); X is the $N_t \times T$ complex data matrix to be transmitted, and D_l is the lth $N_t \times T$ random binary matrix generated randomly where each of its entries is either 0 or 1 and occurs with probabilities P_0 and P_1, respectively. For each l, a new l^{th} D_l is constructed with different positions of zeros. This means that $D_1 \neq D_2 \neq D_l \neq \cdots \neq D_L$ and consequently, $X_1 \neq X_2 \neq X_l \neq \cdots \neq X_L$. Such a method is considered as rateless coding in the sense that the encoder can generate on the fly potentially a very large number of blocks. A power constraint on each X_l is introduced as the average power does not exceed N_t.

Now, we consider a downlink massive multiuser MIMO (MU-MIMO) system in which RSTBC is applied as shown in **Figure 7**.

In this system, a BSTx, equipped with a large number of antennas, communicates simultaneously with K independent users on the same time-frequency resources where each user device has N_r receive antennas. The overall channel matrix $H_a \in \mathscr{C}^{KN_r \times N_t}$ can be written as

$$H_a = \begin{bmatrix} H_1 & H_2 & \cdots & H_K \end{bmatrix}^T \tag{2}$$

where $H_k \in \mathscr{C}^{N_r \times N_t}$, $k = 1, 2, \cdots, K$, is the channel matrix corresponding to the kth user. To eliminate the effects of the multiuser interference (MUI) at the specific receiving users, a precoding technique is applied at the BSTx with, for example, a zero-forcing (ZF) precoding matrix $G_a \in \mathscr{C}^{N_t \times KN_r}$ which is calculated as

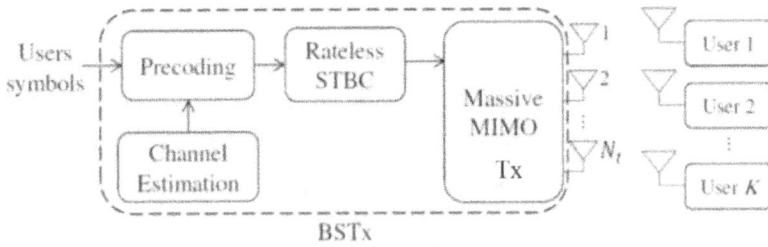

Figure 7. RSTBC code for massive MU-MIMO system (BS-to-users scenario).

$$G_a = \beta \hat{H}_a^H \left(\hat{H}_a \hat{H}_a^H \right)^{-1} \tag{3}$$

where β is a normalized factor.

In this system, channel reciprocity is exploited to estimate the downlink channels via uplink training, as the resulting overhead is linearly a function of the number of users rather than the number of BS antennas [46].

For a single-cell MU-MIMO system, the received signal at the kth user at time instant t can be expressed as

$$y_k = \sqrt{\frac{E_x}{L N_t N_0}} \sum_{l=1}^{L} \sum_{n=1}^{N_t} h_{kn} x_n d_{nl} + w_k \tag{4}$$

where $\frac{E_x}{N_0}$ corresponds to the average SNR per user (E_x is the symbol energy, and N_0 is the noise power at the receiver); L is the maximum number of required blocks of RSTBC at the user; $l = 1, 2, \cdots,$ $L; k = 1, 2, \cdots, K; x_n$ is the transmitted signal by the the nth antenna where $n = 1, 2, \cdots, N_t; h_{kn}$ is the channel coefficient from the the nth transmit antenna to the kth user; $d_{nl} \in \{0, 1\}$ which is the $(n, l)^{th}$ element of the matrix $D;$ and w_k is the noise at the kth user receiver.

It has been demonstrated in [42–45] that RSTBC is able to compensate for data losses. For more details, the reader is referred to these references. Here are some sample simulation results. The averaged symbol-error-rate (SER) performance when RSTBC is applied for $N_t = 100$ with QPSK is shown in **Figure 8**, where the loss rate is assumed to be 25%.

It is inferred from **Figure 8** that for small values of L, the averaged SER approaches a fixed level at high SNR because RSTBC, with the current number of blocks, is no longer able to compensate for further losses. Therefore, it is required to increase L to achieve enhancements until losses effects are eliminated. As shown, for instance, RSTBC with $L = 32$, the flooring in the SER curves has vanished due to the diversity gain achieved by RSTBC (as the slopes of the SER curves increase) so that the effect of losses is eliminated considerably. Thus, the potential for employing RSTBC to combat losses in massive MU-MIMO systems has been shown.

Furthermore, **Figure 9** shows the cumulative distribution function (CDF) of the averaged downlink SINR (in dB) in the target cell for simulation and analytical results for a multi-cell massive MU-MIMO system with $N_t = 100$, $K = 10$ users, QPSK, and pilot reuse factor = 3/7,

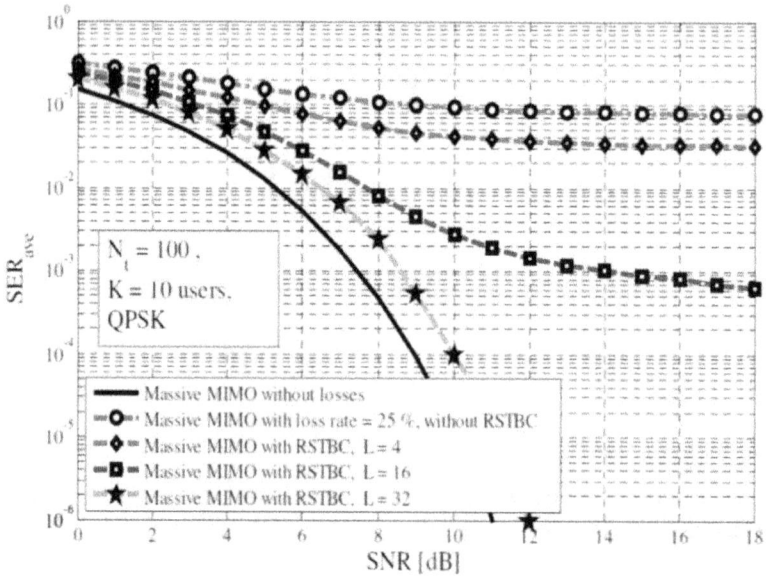

Figure 8. SER curves for massive MU-MIMO system with 25%-rate loss and N_t = 100, K = 10 users, with QPSK, when RSTBC is applied.

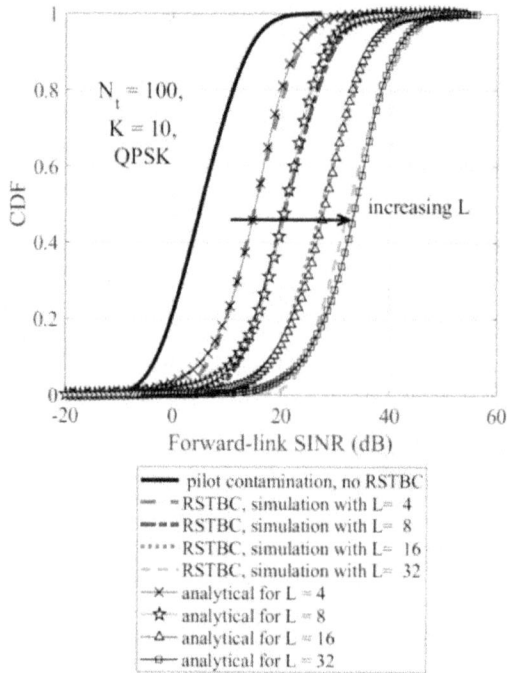

Figure 9. CDF simulation and analytical results' comparisons of SINR for multi-cell massive MU-MIMO system with N_t = 100, K = 10, QPSK, pilot reuse factor = 3/7, RSTBC with L = 4, 8, 16, 32, and 25% loss rate. Analytical curves are plotted using Eq. (21) in [43].

Parameter	Value
Cell radius	500 m
Reference distance from the BS	100 m
Path loss exponent	3.8
Carrier frequency	28 GHz
Shadow fading standard deviation	8 dB

Table 1. Simulation parameters for massive MU-MIMO system.

when RSTBC is applied with L = 4, 8, 16, 32, where lossy channel of 25% loss rate is assumed. Notably, RSTBC supports the system to alleviate the effects of pilot contamination by increasing the downlink SINR. Simulation and analytical results show good matching as seen. Also, it is obvious that the improvements in SINR are linear functions of the number of RSTBC blocks L. It should be mentioned that the simulation parameters are tabulated in **Table 1**.

8. Conclusion

In this chapter, we have considered the rateless space-time block code (RSTBC) for massive MIMO wireless communication systems. Unlike the fixed-rate codes, RSTBC adapts the amount of redundancy over time and space for transmitting a message based on the instantaneous channel conditions. RSTBC can be used to protect data transmission in lossy systems and still guarantee the reliability of the regime when transmitting big data. It is concluded that, using RSTBC with very large MIMO dimensions, it is possible to recover the original data from a certain amount of encoded data even when the losses are high. Moreover, RSTBC can be employed in a multi-cell massive MIMO system at the BS to mitigate the downlink inter-cell interference (resulting from pilot contamination) by improving the downlink SINR. These results strongly introduce the RSTBC for the upcoming 5G wireless communication systems.

Author details

Ali Alqahtani

Address all correspondence to: ahqahtani@ksu.edu.sa

College of Applied Engineering, King Saud University, Riyadh, Saudi Arabia

References

[1] Abdullah A, Abbasi M, Fisal N. Review of rateless-network-coding based packet protection in wireless sensor networks. Mobile Information Systems. 2015;**2015**:1-13

[2]   Liew T, Hanzo L. Space–time codes and concatenated channel codes for wireless communications. Proceedings of the IEEE. 2002;**90**(2):187-219

[3]   Huang J-W, Yang K-C, Hsieh H-Y, Wang J-S. Transmission control for fast recovery of rateless codes. International Journal of Advanced Computer Science and Applications (IJACSA). 2013;**4**(3):26-30

[4]   Bonello N, Yang Y, Aissa S, Hanzo L. Myths and realities of rateless coding. IEEE Communications Magazine. 2011;**49**(8):143-151

[5]   Bonello N, Zhang R, Chen S, Hanzo L. Reconfigurable rateless codes. In: IEEE 69th Vehicular Technology Conference, 2009, VTC Spring 2009; IEEE. 2009, pp. 1-5

[6]   Bernard S. Digital Communications Fundamentals and Applications. USA: Prentice Hall; 2001

[7]   Mehran F, Nikitopoulos K, Xiao P, Chen Q. Rateless wireless systems: Gains, approaches, and challenges. In: 2015 IEEE China Summit and International Conference on Signal and Information Processing (Chi-naSIP). IEEE; 2015. pp. 751-755

[8]   Wang X, Chen W, Cao Z. ARQ versus rateless coding: from a point of view of redundancy. In: 2012 IEEE International Conference on Communications (ICC); IEEE. 2012. pp. 3931-3935

[9]   Wang P. Finite length analysis of rateless codes and their application in wireless networks [PhD dissertation]. University of Sydney; 2015

[10] Byers JW, Luby M, Mitzenmacher M, Rege A. A digital fountain approach to reliable distribution of bulk data. ACM SIGCOMM Computer Communication Review. 1998;**28**(4):56-67

[11] Luby M. LT codes. In: The 43rd Annual IEEE Symposium on Foundations of Computer Science, 2002. Proceedings. 2002. pp. 271-280

[12] Shokrollahi A. Raptor codes. IEEE Transactions on Information Theory. 2006;**52**(6):2551-2567

[13] Maymounkov P. Online codes. Technical report. New York University; 2002

[14] Palanki R, Yedidia JS. Rateless codes on noisy channels. In: International Symposium on Information Theory, 2004. ISIT 2004. Proceedings; June 2004; p. 38

[15] Chong KFE, Kurniawan E, Sun S, Yen K. Fountain codes with varying probability distributions. In: 2010 6th International Symposium on Turbo Codes Iterative Information Processing; Sept 2010. pp. 176-180

[16] Kuo S-H, Lee H-C, Ueng Y-L, Lin M-C. A construction of physical layer systematic Raptor codes based on protographs. IEEE Communications Letters. 2015;**19**(9):1476-1479

[17] Chen S, Zhang Z, Zhu L, Wu K, Chen X. Accumulate rateless codes and their performances over additive white Gaussian noise channel. IET Communications. March 2013; 7(4):372-381

[18] Erez U, Trott MD, Wornell GW. Rateless coding for Gaussian channels. IEEE Transactions on Information Theory. Feb 2012;58(2):530-547

[19] Castura J, Mao Y. Rateless coding over fading channels. IEEE Communications Letters. Jan 2006;10(1):46-48

[20] Castura J, Mao Y. Rateless coding and relay networks. IEEE Signal Processing Magazine. Sept 2007;24(5):27-35

[21] Tian S, Li Y, Shirvanimoghaddam M, Vucetic B. A physical-layer rateless code for wireless channels. IEEE Transactions on Communications. June 2013;61(6):2117-2127

[22] Shanechi MM, Erez U, Wornell GW. Rateless codes for MIMO channels. In: IEEE GLOBECOM 2008-2008 IEEE Global Telecommunications Conference; Nov 2008. pp. 1-5

[23] Wolniansky PW, Foschini GJ, Golden G, Valenzuela RA. V-BLAST: An architecture for realizing very high data rates over the rich-scattering wireless channel. In: 1998 URSI International Symposium on Signals, Systems, and Electronics, 1998. ISSSE 98; IEEE. 1998. pp. 295-300

[24] Fan Y, Lai L, Erkip E, Poor HV. Rateless coding for MIMO fading channels: performance limits and code construction. IEEE Transactions on Wireless Communications. 2010; 9(4):1288-1292

[25] Zheng L, Tse DNC. Diversity and multiplexing: A fundamental trade-off in multiple-antenna channels. IEEE Transactions on Information Theory. 2003;49(5):1073-1096

[26] Tavildar S, Viswanath P. Approximately universal codes over slow-fading channels. IEEE Transactions on Information Theory. 2006;52(7):3233-3258

[27] Chen Y, Huang H, Zhang Z, Qiu P, Lau VK. Cooperative spectrum access for cognitive radio network employing rateless code. In: ICC Workshops-2008 IEEE International Conference on Communications Workshops; IEEE. 2008. pp. 326-331

[28] Chen X, Zhang Z, Chen S, Wang C. Adaptive mode selection for multiuser MIMO downlink employing rateless codes with QoS provisioning. IEEE Transactions on Wireless Communications. 2012;11(2):790-799

[29] Afzal J, Stockhammer T, Gasiba T, Xu W. System design options for video broadcasting over wireless networks. In: Proceedings of IEEE CCNC, vol. 54. Citeseer; 2006. p. 92

[30] Afzal J, Stockhammer T, Gasiba T, Xu W. Video streaming over MBMS: A system design approach. Journal of Multimedia. 2006;1(5):25-35

[31] Molisch AF. Wireless Communications, vol. 2. New York, USA: John Wiley & Sons; 2011

[32] Labiod H. Wireless ad hoc and Sensor Networks. Vol. 6. New York, USA: John Wiley & Sons; 2010

[33] Hartung F, Horn U, Huschke J, Kampmann M, Lohmar T, Lundevall M. Delivery of broadcast services in 3G networks. IEEE Transactions on Broadcasting. 2007;53(1):188-199

[34] Culler D, Estrin D, Srivastava M. Guest editors' introduction: Overview of sensor networks. IEEE Computer Society. Aug 2004;37(8):41-49

[35]   Zhao F, Guibas LJ. Wireless Sensor Networks: An Information Processing Approach. San Francisco, USA: Elsevier Science & Technology; 2004

[36]   Razavi R, Fleury M, Ghanbari M. Rateless coding on a wearable wireless network for augmented reality and biosensors. In: 2008 IEEE 19th International Symposium on Personal, Indoor and Mobile Radio Communications; IEEE. 2008. pp. 1-4

[37]   Erez U, Wornell G, Trott MD. Rateless space–time coding. In: Proceedings. International Symposium on Information Theory, 2005. ISIT 2005; IEEE. 2005, pp. 1937-1941

[38]   Wang C, Zhang Z. Performance analysis of a rate varying space–time coding scheme. In: 2013 International Workshop on High Mobility Wireless Communications (HMWC). IEEE. 2013. pp. 151-156

[39]   Larsson E, Edfors O, Tufvesson F, Marzetta T. Massive MIMO for next generation wireless systems. Communications Magazine, IEEE. 2014;**52**(2):186-195

[40]   Tarokh V, Seshadri N, Calderbank AR. Space–time codes for high data rate wireless communication: Performance criterion and code construction. IEEE Transactions on Information Theory. 1998;**44**(2):744-765

[41]   Alamouti SM. A simple transmit diversity technique for wireless communications. IEEE Journal on Selected Areas in Communications. 1998;**16**(8):1451-1458

[42]   Alqahtani AH, Sulyman AI, Alsanie A. Rateless space time block code for massive MIMO systems. International Journal of Antennas and Propagation. 2014;**2014**:1-10

[43]   Alqahtani AH, Sulyman AI, Alsanie A. Rateless space time block code for mitigating pilot contamination effects in multicell massive MIMO system with lossy links. IET Communications Journal. 2016;**10**(16):2252-2259

[44]   Alqahtani AH, Sulyman AI, Alsanie A. Rateless space time block code for antenna failure in massive MU-MIMO systems. IEEE Wireless Communications and Networking Conference (WCNC); Doha, Qatar; April 2016. pp. 1-6

[45]   Alqahtani AH, Sulyman AI, Alsanie A. Loss-tolerant large-scale MU-MIMO system with rateless space time block code. In: 22nd Asia-Pacific Conference on Communications (APCC); Yogyakarta, Indonesia; August 2016. pp. 342-347

[46]   Marzetta TL. How much training is required for multiuser MIMO? In: Fortieth Asilomar Conference on Signals, Systems and Computers, 2006. ACSSC'06; IEEE; 2006. pp. 359-363

Chapter 4

Evolution and Move toward Fifth-Generation Antenna

Kioumars Pedram, Mohsen Karamirad and
Negin Pouyanfar

Additional information is available at the end of the chapter

http://dx.doi.org/10.5772/intechopen.74554

Abstract

With the introduction of various antennas in the field of antenna technology, most of the constraints related to the transmission and receiving of the signals at different intervals have been resolved. By the rapid growth in industry and consequently high demands in the communication arena, the conventional antennas are unable to respond to these extended requirements. However, those initial antennas were suitably used in the field of technology. In the recent decades, by introducing new antenna technologies such as metamaterial structures, substrate integrated waveguide (SIW) structures and microstrip antennas with various feeding networks could meet the demands of the current systems. As stated before, in the frequency ranges of below 30 GHz, antenna size and bandwidth are of the important issues, so that novel antennas can be created in low frequencies, which are able to achieve reliable radiation properties when combined with new multiband antennas. Generally, transmission lines are practical in low frequencies and short distances, while higher frequencies are mainly used due to bandwidth goals. This chapter is organized into three subsections related to the 5G wireless communication systems: antennas below 15 GHz or accordingly antennas with wavelength less than 1/20; antennas operating between 15 and 30 GHz; higher frequency antennas or millimeter-wave antennas, which are desired for above 40 GHz.

Keywords: multiband antenna, 5G application, mm-wave, substrate integrated waveguide (SIW), MIMO, array antenna

1. Introduction

From the beginning of human civilization, communications had basic importance for human society in recent years' human used electromagnetic spectrum beyond of visible area for telematics communication through radio waves radio antenna is a basic part of a radio system. A radio antenna is a tool that provides the possibility of radiation or receiving radio waves. As we know, one of the biggest human sources is electromagnetic spectrum and antennas played a basic role in using this natural source. Despite several antennas in techs many of limitations have been solved in sending and receiving in some areas. The first practical cellular network which used analog systems arose in 1982 as a first generation in 1991 the analog system had been improved to the digital one or internet-based generation named as second generation this technology also added cellular data in format of General Packet Radio Service (GPRS) and Enhanced Data rates for GSM Evolution (EDGE). Approximately 10 years later, the third generation had been introduced to improve data rate further. After a decade, the current LTE networks had emerged which commonly named fourth generation. A new generation of network tech has been used the fifth generation of network or 5G like previous generation mutation will improve in cases like a speed velocity fifth generation. Some believe that new generation of mobile networks received new frequencies bands braid band received wide spectrum in each frequencies channel. Some believe that the new mobile generations usually provide new and wider frequency bands compared to the first, second, third and fourth generations, which provide up to 30 kHz, 200 kHz, 5 MHz and 20 MHz, respectively. The higher frequency band may interfere with K band which is specified for satellite communications. From user's perspectives, the former mobile generation (4G) provides a significant peak rate bit up to 1Gbps. Therefore, for the 5G to be different, it should support more data rate speed, so that makes it possible to connect different devices, simultaneously along with higher spectrum efficiency (more data rate per unit), low battery usage, lower delay, and disconnection. Moreover, lower cost for establishment of infrastructures, flexibility, higher scalability, and reliability should be taken into consideration. The existing problem is that higher radio frequencies cannot perform well in long distances or cross the walls between them and mobile devices. Thus to solve the aforementioned problems, service providers must focus on the new communication antenna technologies. Antennas with high input and output capacities will be able to transmit parallel radio waves which accordingly define a signal beam and eventually the radio signal energy directs toward a specific path, which the user is situated. Since antennas are key elements of wireless communication systems, an expert design can meet the demands of systems and consequently improve system performance. Antenna role in the communication systems is similar to the eyes and glasses for a human. Antennas scope of activities is extensive and dynamic so that during the past six decades, antenna technology has had an undeniable evolution in the communication arena. Most of the important improvements in this field are currently used by public users. However, today we face more challenges since the system efficiencies have been noticed. Most of these improvements in antenna technology have been evolved since 1970.

2. Antenna in 5G application

The initial research in the field of fifth generation has been started since 2012. With the standardization procedure of this generation of the mobile telecommunication networks, beginning since 2015, it is anticipated that first experimental samples will be set up in 2018. Based on the most forecasts, commercialization of these networks will be postponed to the next 2 or 3 years. Many investigations have been done in this field in the past few years so that some researchers have worked on antennas to improve their impedance bandwidth. In some cases, unidirectional pattern concentration with high gain or pattern rotating has been tried. Recently due to the tremendous increase in the number of devices connected to the wireless communication systems and accordingly a significant increase in demands for new and high-quality applications, antennas with wider impedance bandwidth, high gain and rotatable radiation pattern especially in higher frequencies are required.

2.1. Antennas below 15 GHz or accordingly antennas with a wavelength less than 1/20

Radio frequency has a set of physical properties. One of those is the wavelength of the signal. At 2.4 GHz this is approximately 12.5 cm (4.92 inches) and 5–6 cm (2–2.3 inches) at 5 GHz as well as 2.5–3 cm (1–1.15 inches) at 10 GHz. The difference and approximation are due to the fact that the wavelength is the result of the direct correlation of the exact frequency (2.400–2.483.5 GHz in the 2.4 GHz range and 5.250–5.725 GHz in the 5 GHz range). To optimize sending and receiving the signal, the antenna is designed around those physical properties. The elements inside the antennas will vary in size to match the wavelength (or more commonly 1/4 or 1/8 or 1/16th the size of the wavelength). So first and the foremost difference in between is the size of the antennas. The 2.4 GHz antennas are bigger than the 5 GHz antennas. Mind that the same size antenna enclosure may be used for various reasons, two biggest ones being the cost of development and production and also overall esthetics. There are many antenna types available: dipole omni antennas, patch, and Yagi antennas, just to name a few. There are many subtypes, too many to name all of them here. Different antenna type will provide different radiation pattern. Starting with Dipole-Omni antenna that will provide 360° coverage in vertical setup (point of the antenna facing straight up or down) to focus, narrow beamwidth antennas used for Point to Point communication and everything in between. The RF focus will result in the higher gain of the antenna as it directs all the available energy into a certain direction [1–3].

2.1.1. Antenna based on SIW structure for 5G application

A Substrate Integrated Magneto-Electric Dipole antenna has been introduced for 5G Wi-Fi applications in [1]. A new technique is used to reduce the height of the ME dipole antenna by utilizing the tapered H-shape ground plane. As the conventional approach of folding parallel walls is hard to fabricate and also challenging, a pair of open slots has been cut on the ground plane. Hence surface current path is folded along the x-axis. The proposed configuration consists of four layers as shown in **Figure 1**. The antenna is fed by a Γ-shaped probe located between the two arms of the Bowtie dipole. The simulated and measured results shown in **Figure 2** reveal that the proposed structure provides an impedance bandwidth of 18.74% between 4.98 and 6.01 GHz and gains of about 6.8 and 7.2 dBi, respectively.

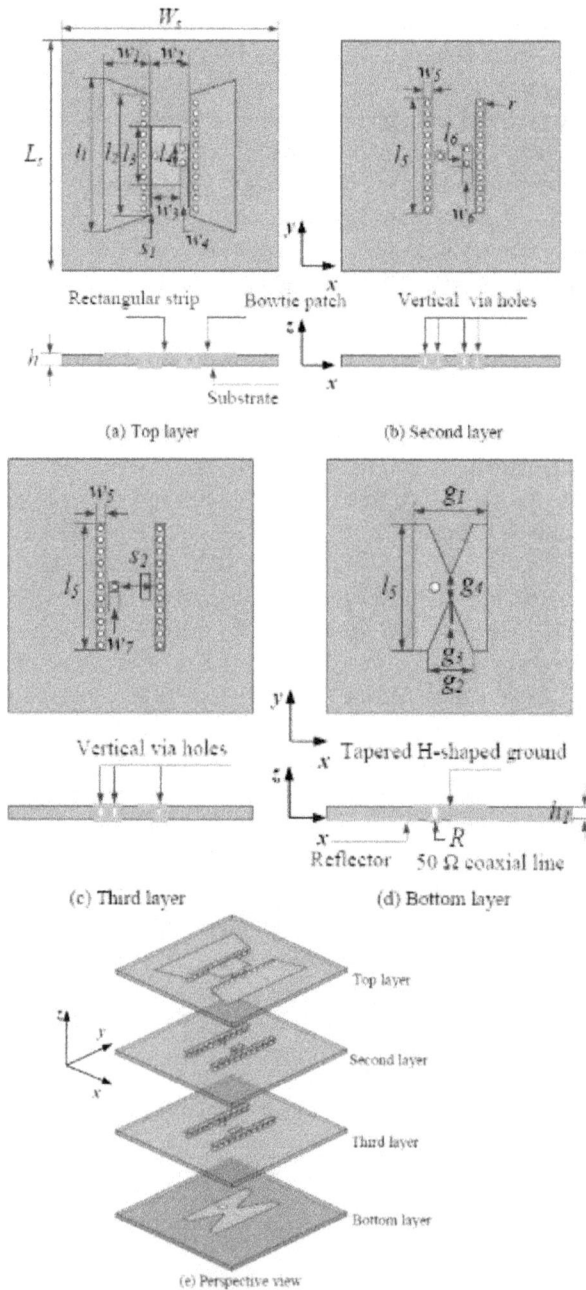

Figure 1. Geometry of the substrate integrated ME-dipole [1].

With the outstanding features with respect to the conventional ones, including wide impedance bandwidth, symmetric patterns, low back radiation and at the same time more than
80% efficiency, the proposed multilayer configuration with a novel technique to reduce the
height of the ME dipole antenna is a suitable candidate for 5G Wi-Fi application.

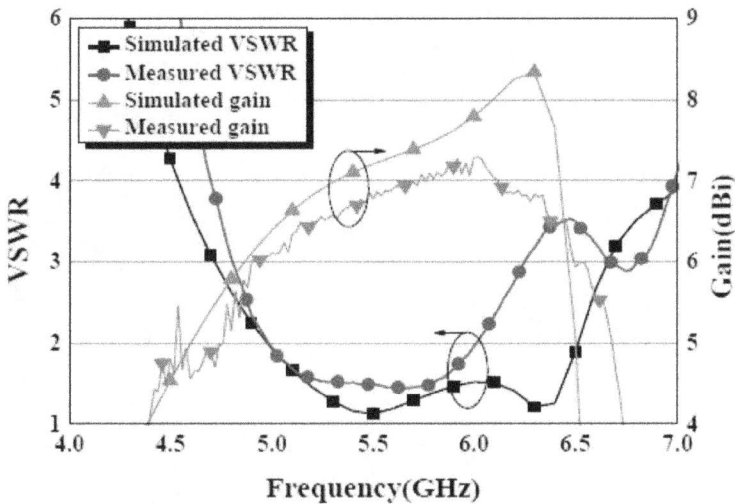

Figure 2. Simulated and measured VSWR and gains versus frequency [1].

A new configuration based on substrate integrated waveguide concept has been investigated for 10 GHz frequency band in [2]. In order to enhance the antenna gain, different structures with two and four grooves have been proposed. The aperture-coupled antenna consists of a rectangular patch in which the power is coupled through the cavity and hence into free space. The cavity dimensions are optimized so that they can determine resonance frequency and radiation pattern of the antenna. The geometry of the designed configurations with two and four grooves and their corresponding simulated results are shown in **Figure 3**. Via based cavity structure has a remarkable impact on cavity performance in SIW patterns and it is demanding to adjust its resonance frequency due to abrupt changes of current distributions near walls. Therefore, the rectangular cavity's edges are created by drilling out the two L-shapes in the middle substrate. Despite of drilling out the shapes, the corners have been left to keep the substrate inside the cavity in place. Thereafter, the walls of the cavity are metallized as a PCB manufacturing process.

The simulated reflection coefficient reveals that adding the grooves have approximately no effect on the S11 if the distance between patch and grooves are accurately determined. Moreover, this distance changes the directionality of the antenna, as depicted in **Figure 4**.

2.2. Antenna in mid-frequency (15 and 30 GHz) for 5G application

Due to the steep increase in the number of electronic devices and accordingly high data traffic, wireless communication technology is required to use higher frequency bands to overcome the shortcomings existed in the existing networks. Fifth generation (5G) mobile networks have been extremely noted to overcome the existing networks problems such as bandwidth shortage caused as a result of the exponentially growth in the number of electronic devices and users connected to wireless systems, since it can provide a peak data rate of at least 100 Mb/s in urban areas, 10 Gbps for static users and 1 Gbps for mobile users. Two of the important bands specified for testing 5G cellular communication systems are 28 and 39 GHz in the US and Europe, respectively. These frequency ranges extend from 24.5 to 29.5 GHz

(a)   (b)

(c)

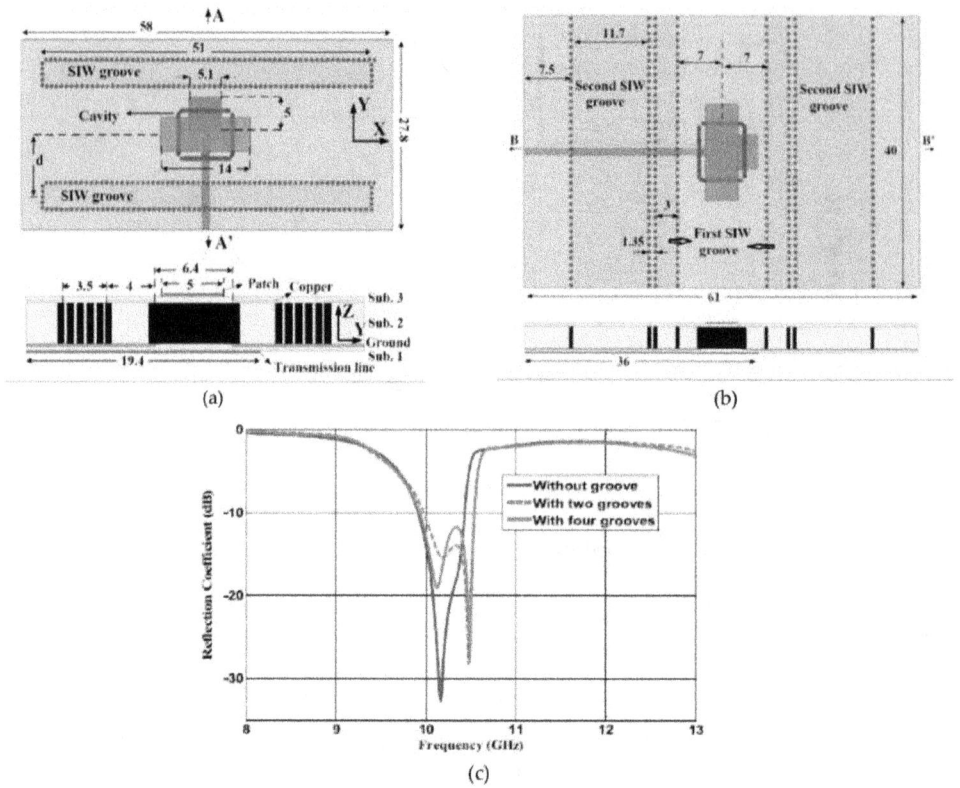

Figure 3. Configuration of proposed aperture antenna (a) with two grooves, (b) with four grooves, (c) reflection coefficient of the proposed antenna without/with grooves [2].

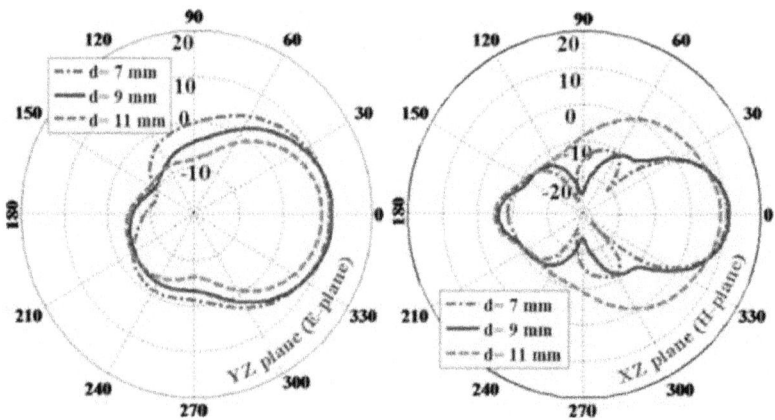

Figure 4. Radiation pattern of the proposed antenna with two grooves for different distance between patch and grooves for center frequency [2].

and 37.0 to 43.5 GHz with center frequencies of 28 and 39 GHz, respectively. Improved data rates of up to 2.5 Gbps with multiple connections are among important characteristics of 5G cellular communications [4]. The US Federal Communications Commission (FCC) has recommended frequency bands of 28 and 37–39 GHz for the ongoing wireless networks (5G), as well as 33 GHz, which was specified for satellite and navigation applications. The Ka-band (at 28/38GHz) can be suitable for frequency division multiplexing (FDD), in which single antennas providing dual-band performance is preferred. The significantly increased path loss at very high-frequencies has to be compensated by higher antenna gains, which is made possible by increasing the number of antennas at the base station [5]. Compared to the current 1–2 GHz cellular bands, the spectrum at 28 GHz has less free space path loss. In fact, oxygen loss (due to oxygen molecule absorption in atmosphere) and rain attenuation will have less impact on the 28 GHz spectrum, hence providing better propagation conditions when compared to the existing cellular networks. It should be noted that the 28/38 GHz signals will not be going to penetrate a car's windows or roof. Therefore, these kinds of devices will be important for these frequency bands in direct communication with a user device. For future 5G applications, it has been indicated in [6] that, a high gain antenna (>12 dB) is required, which has the capability to be directed in certain directions. One introduced technology to overcome the existing deficiencies and to meet the aforementioned goals is the massive MIMO, which means extending MIMO concept to hundreds of antennas at the base station as a promising solution to increase data rate and network capacity by allowing beamformed data [7]. Another introduced technology to obtain an efficient beam steering characteristic is to use phased array antennas, which are one of the key parts in 5G wireless systems, since smaller antennas can be employed as arrays to improve performance [8]. It should be noted that the capability of beam steering in antennas is not compatible with most of the previous generations (2G, 3G and 4G), since they usually broadcast signals in wide beams, hence dissipating energy in unwanted directions. The importance of utilizing phased array antennas is that they can direct and accordingly focus the signal beams to a desired direction toward the receiving antenna. As an example, the IBM and Ericsson has designed a phased array, which supports beam steering of less than 1.4° for focusing the beam toward users. There have been some techniques to design multiband antennas, among which is the slotted-SIW structures. The slotted SIW is a good option for designing the directional multi-band antennas. By utilizing different slot configurations in these antennas more directional radiation patterns can be obtained. This can be explained such that in the SIW structures usually one of the layers contains the ground and the other has the radiating apertures. The surface current is disturbed by the engraved slots to accurately radiate electromagnetic waves. This method can also be used for the incoming wireless network. As the modern wireless systems require low profile and easy to integrate devices providing high gain and efficiency, microstrip patch antennas are highly recommended. Therefore, small antennas with dual-band or multiband properties are preferred for applications operating in Ka-band for the futuristic 5G technology.

A microstrip array antenna designed in [9] provides an impedance bandwidth ($S_{11} < -10$ dB) of about 7 GHz from 23.9 to 31 GHz and 12.5 dB gain at 29 GHz, but poor radiation efficiency due to the lossy FR-4 substrate has been observed. Since the scanning angle of more than 45° is required for mobile antennas, a planar array with beam switching capability is proposed in [10], which provides 1 GHz impedance bandwidth relative to the center frequency of 28.05 GHz and an average of 10 dBi gain. Another 28 GHz multilayer FR-4 PCB antenna array is presented in [11]. The structure consists of 16-element shows a fan-beam like radiation pattern. Results indicate that nearly 11 dB gain and more than 3 GHz impedance bandwidth has been obtained. The presented array is a complicated and expensive structure due to the multilayer technology. Some recently published configuration to achieve the acceptable performance for these important frequencies will be discussed in detail. Three different oriented microstrip inset fed patches are designed in [12]. In the first design, two patches are placed side by side while in the other two configurations opposite feeding structures are used. The geometry of all structures and their corresponding reflection coefficients are presented in **Figures 5** and **6**, respectively. It can be seen in all three configurations, S_{11} is the same and about 1.5 GHz, which is due to the symmetrical structures of the antennas. The simulated and measure mutual coupling for structure 3 is shown in **Figure 6**. As it is obvious because of the small size of the antenna S12 < −20 dB at 28 GHz has been obtained.

A compact, broadband printed-dipole antenna, and its' corresponding 8-element array antenna have been investigated in [10] to work at 28/38 GHz, which are the key frequency band for ongoing 5G. The single element design consists of microstrip feed line on the top layer of the substrate and the dipole along with the ground plane on the other layer. An integrated balun including a 45° folded microstrip and a rectangular slot has been employed and optimally adjusted to improve impedance matching. The bandwidth of about 36% in the frequency range of 26.5–38 GHz and 4.5–5.8 dBi gain has been obtained. The simulated and measured results for the single element proposed antenna are shown in **Figure 7**.

2.3. Higher frequency antennas or millimeter-wave antennas which are desired for 60 GHz

With recent ongoing advances in new generations of telecommunications, the research on 60 GHz antenna design has become progressive, since their ability to provide high data rate services for fifth-generation (5G) applications. As a matter of fact, the implantation of 5G networks requires wide bandwidth which satisfies the demand to have real-time video streaming, machine to machine communications and IOT. For the sake of aforesaid, providing broadband infrastructure is a noticeable challenge in 60 GHz technology as it is an alternative of fiber optics. Thereby, the design of antenna with a low profile, high gain, and high radiation efficiency is necessary. Hitherto, some efforts have been conducted to alleviate these requirements [13].

2.3.1. Wideband linearly polarized transmitarray antenna for 60 GHz

In this section, a transmitarray antenna for backhauling at V-band is discussed. Such high capacity is of interest for operators to have multihop in the ranges of hundreds of meters to 1 km. for this purpose, three different frequency bands (28 GHz, V-band, and E-band) are dedicated to millimeter wave backhauling. According to the substrate integrated waveguide-based planar array have been undertaken. Contrary to microstrip based arrays in which high insertion loss hampers the performance, the employment of spatial feeding illumination in

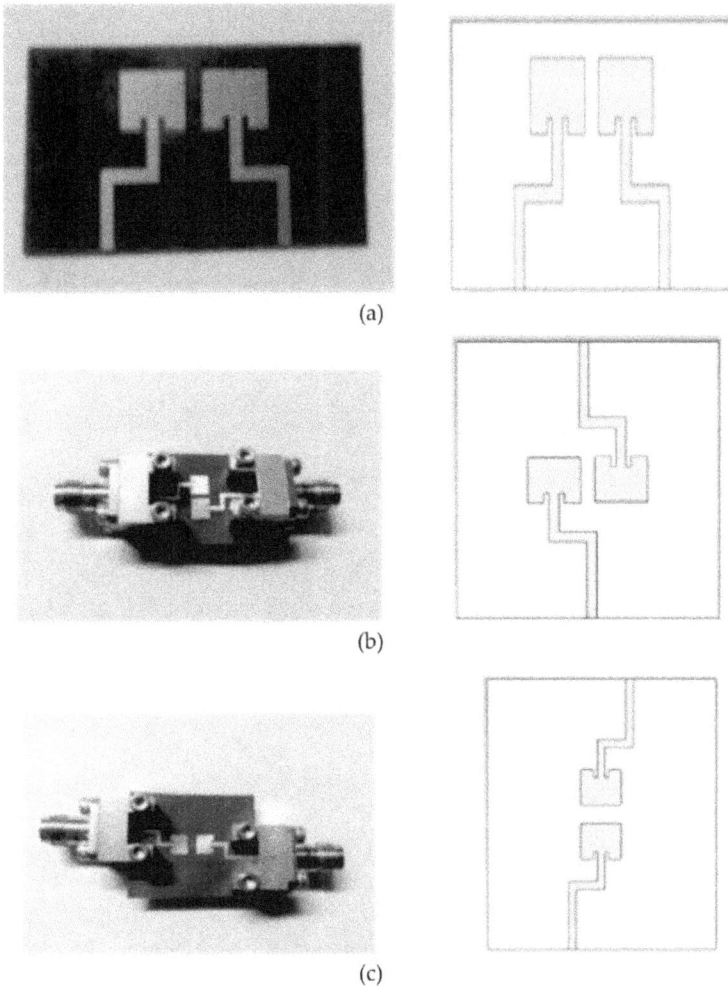

Figure 5. Three different configurations of patch array antennas designed. (a) Antenna 1; (b) antenna 2; (c) antenna 3 [9].

transmitarray antenna has drawn considerable research interest. While one or more focal sources illuminate, each unit cell of the transmitarray as a concept have been made of Rx antenna coupled to Tx antenna. Realizing the transmission phase shifting can be obtained by connecting two antennas through a phase shifter. Several studies have been conducted on the transmitarray antenna in last decades, typically with the focus on implementing such patterns in the structures. Recent years, such topological patterns have been demonstrated. One of these is illustrated in **Figure 8** [14].

The proposed structure uses 3-b phase optimization including 0°, 45°, 90°, 135°, 180°, 225°, 270° and 315° over the whole 57–66 GHz. The arrangement of this 3-b is according to two different patterns. The first one is composed of the simple patch which is described in [15]. The

Reflection Coefficient Plot

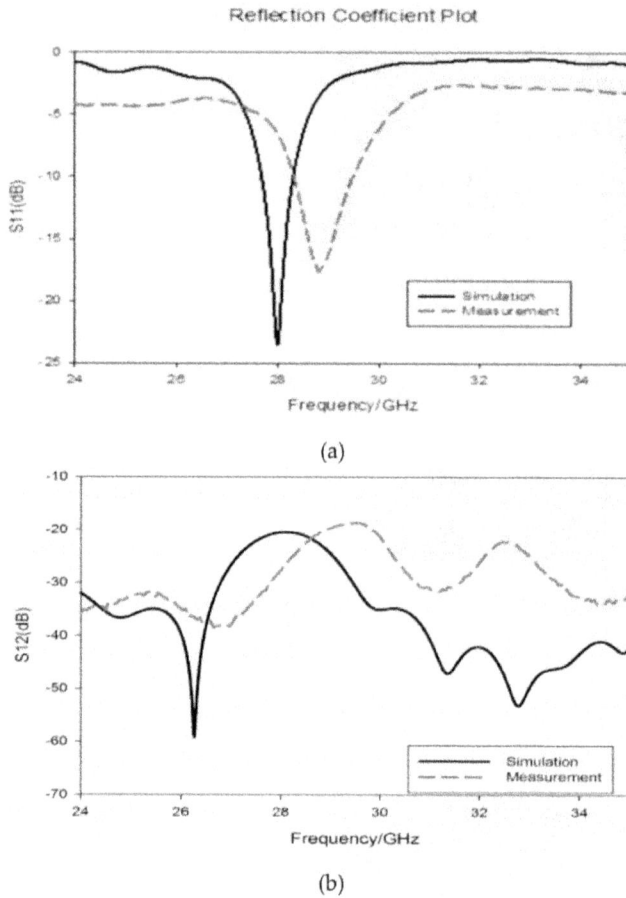

Figure 6. Measured and simulated: (a) return loss for designed antennas 1–3 and (b) mutual coupling for the proposed antenna [9].

second one utilized a capacitive fed patch in the structure. From **Figure 9**, it can be seen that the profiled skirt protecting the antenna from its surrounding, is roughly more complex than standard pyramidal horn structures. Thereupon, a radome is included to protect the array from the environment. The optimization of the unit cell's phase distribution is outperformed using in-house software [16] to gain the radiation pattern [17].

2.3.1.1. Unit-cell design and frequency response

As shown in **Figure 9**, the two types (1 and 2) are demonstrated for the pattern in order to obtain a 3-b phase resolution. Thereafter, using eight unit cell configuration architectures have offered a 45° relative phase shift between each phase state. The unit cell size is optimized to $0.51\lambda_0 \times 0.51\lambda_0$ at 61.5 GHz. The structure is composed of two patch antenna separated with a 508 μm thick dielectric substrate with $\varepsilon_r = 2.5$ and $tan\delta = 0.0017$. The via hole has been utilized

(a)

(b)

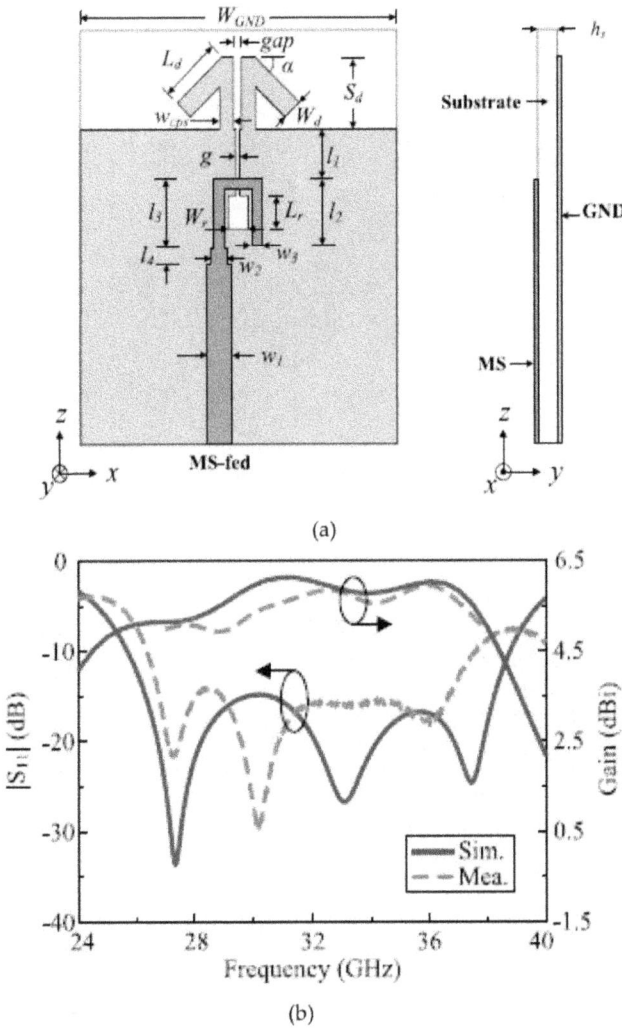

Figure 7. (a) Geometry of the printed-dipole antenna, (b) the simulation and measurement gain and S_{11} of the single element [10].

to ensures the coupling between patches. The configuration of the design is as follows. The unit cell type 1 is designed to reveal the phase state 0° and 45°. It should be noted that the 45° phase shift is derived by resizing the 0° phase state unit cell [18]. In the same way, the unit cell type 2 is used to generate 90° and 135° phase shift.

Figure 10 indicates that the coefficients are lower −9.65 dB over the whole 58–66 GHz. Hence, the ripple inside the presented tapered focal source could be reduced through minimizing the losses of the transmitarray reflection coefficient. Furthermore, the simulated transmission coefficient in **Figure 10(b)** remains approximately better than −1 dB for each unit cell.

Figure 8. Perspective view of the complete transmitarray antenna [14].

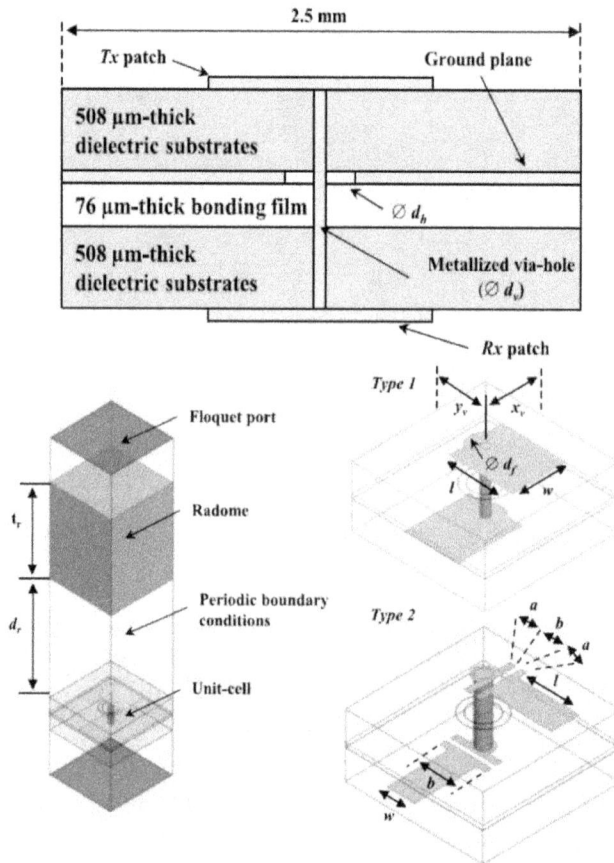

Figure 9. Unit cell cross sectional, detailed pattern and electromagnetic simulation setup with random [14].

At last, the transmission phase is presented in **Figure 10(c)** which illustrates the phase shift of 45° between each unit cell phase resulting in a 3-b phase quantization. As a matter of fact, the transmission coefficient and phases are roughly stabled up to 40° under oblique incident wave as well as investigation in [18, 19].

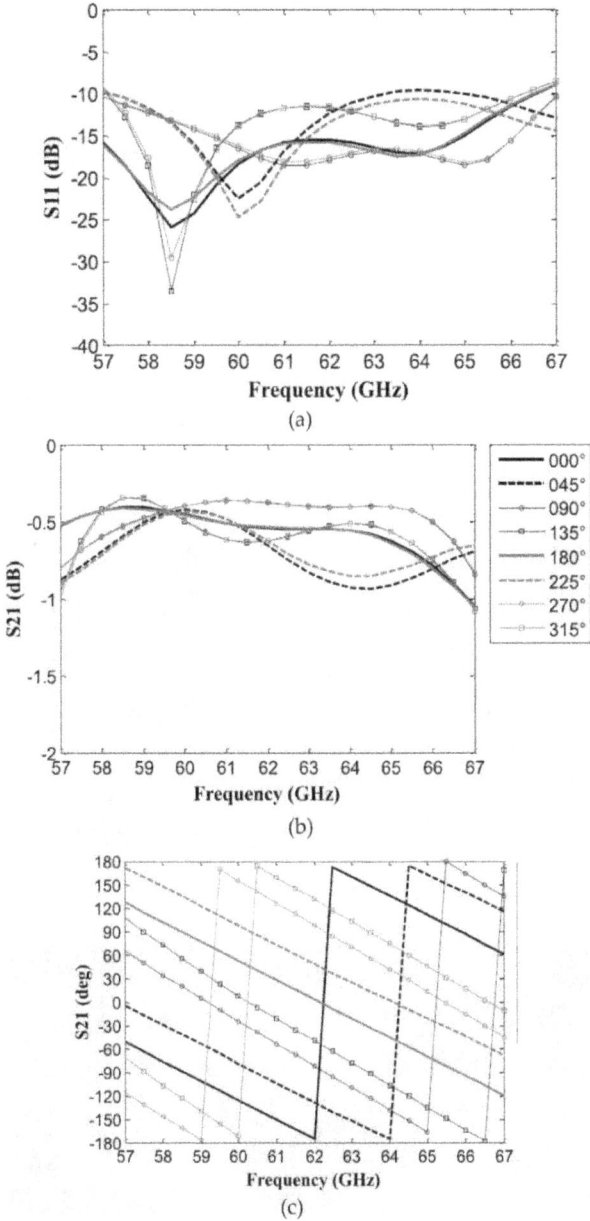

Figure 10. Amplitude of the (a) reflection, (b) transmission coefficient and (c) transmission phase of the unit cell [14].

2.3.1.2. Design and characterization

The simulated frequency response of the transmitarray structure is shown in **Figure 11** for three different array diameters (D). Obviously, it can be found that by reducing the diameter, the maximum gain is reduced, as well. In addition, the transmitarray frequency response is

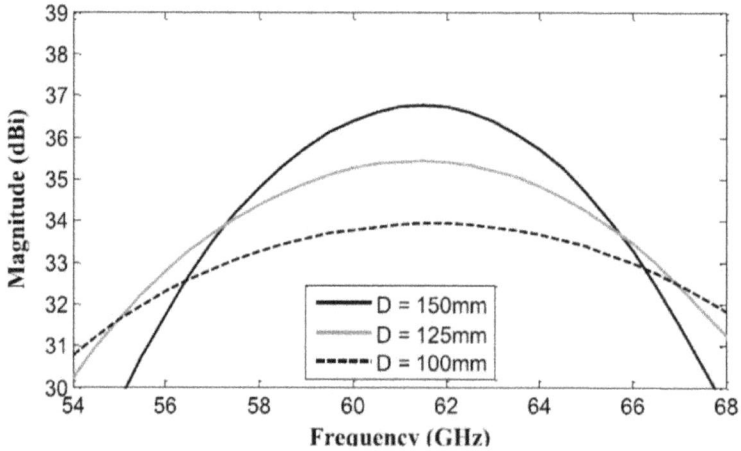

Figure 11. Transmitarray frequency response as a function of the array diameter using the method described in [19].

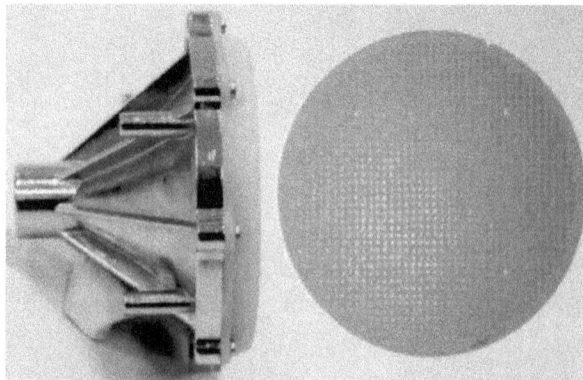

(a)

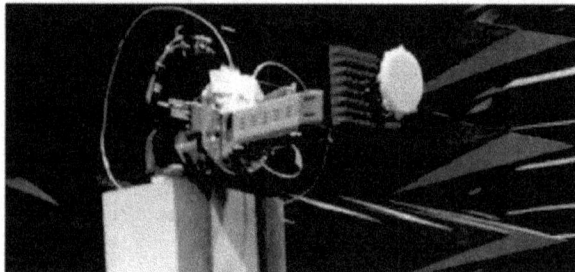

(b)

Figure 12. (a) Fabricated antenna and (b) the transmit array antenna in chamber room during measurement process [14].

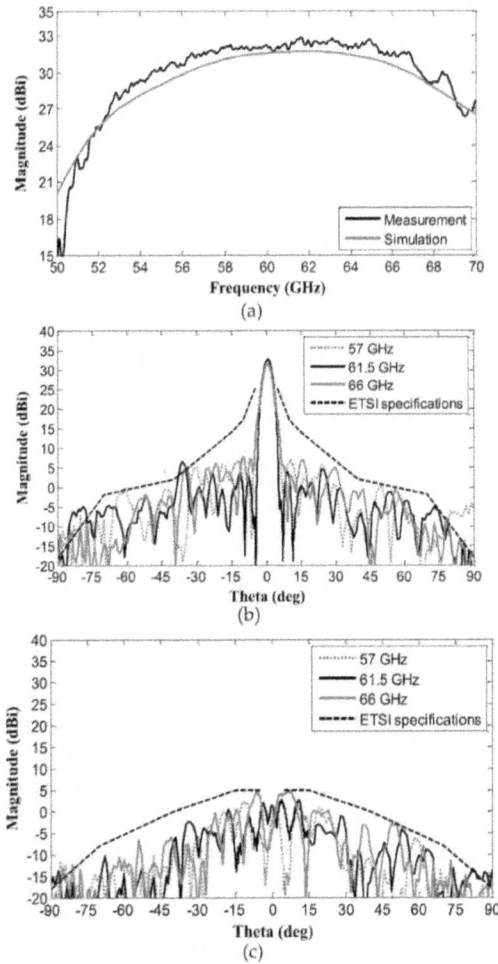

Figure 13. (a) Measured and simulated broadside gain; measured radiation pattern of the linearly polarized transmitarray in (b) co-polar and (c) cross-polar situations [17].

approximately flattened across a stabled gain over the bandwidth of 57–66 GHz. The stabled frequency response can be justified through discussing the phase shift behavior of the unit cell and calculated phase compensation at central frequency 61.5 GHz. In fact, the phase error which is defined as the difference between the calculated phase compensation at a central frequency and at a frequency different from the optimization one increases and a reduction in gain is generated. According to the gain reduction, the true-time-delay technique can be performed which is demonstrated in [20].

2.3.1.3. Fabricated transmit array and experimental results

Figure 12 presents the fabricated 100 mm diameter linearly polarized transmitarray antenna. In order to verify the simulation procedure, **Figure 13(a)** shows the maximum gain in the broadside direction. It can be found that a maximum gain of 32.5 dBi has been obtained with a bandwidth of 15.4% (57–66.5 GHz) through an aperture efficiency of 42.7%. The results of the simulation and measurement are roughly the same, but there are some differences, which have been undertaken. The first one is considering an infinite array of identical elements in simulation procedure. The second is about the non-constant distance between the radome and the planar array, which confirms the mechanical constraints. The measured gain radiation pattern (H-plane) for co- and cross-polar is presented in **Figure 13(b)** and **(c)**. It can be seen that cross polarization discrimination higher than 31 dB has been obtained at three different frequencies (57, 61.5 and 66 GHz).

Although the designed linearly polarized transmit array antenna has been characterized in V-band, a tradeoff between aperture efficiency and array sized is done to guarantee the bandwidth. For this purpose, the designed structure presents a broadside gain of 32.5 dBi at 61.5 GHz with an aperture efficiency of 42.7%. Moreover, the fractional bandwidth of 15.4% from 57 to 66.5 GHz has been obtained across the proposed gain.

3. Conclusion

Many technologies have been introduced and developed for the fifth-generation networks. They are either the evolved shapes of the previous generations technologies or sometimes new. In this chapter we first explained different performance, application and types of antennas used in various frequency bands of fifth generation cellular networks and some works have been discussed in each section. Antennas in frequency bands of less than 15 GHz, which equal to less than 1/20 wavelength, uses new structures to improve bandwidth, beam rotation and energy concentration as well. But for antennas in frequency range of 15–30 GHz and upper than 40 GHz or millimeter-wave antennas, the main issue is the appropriate size of antenna while achieving a good radiation performance. Therefore, in these frequencies horn antennas, MIMO antennas, dielectric resonator antennas and phased array antennas are widely used, so that a reasonable bandwidth can be obtained by beam rotating characteristic of the mentioned antennas. Moreover, by making changes on these antennas or by combining different types, it would be possible to enhance antenna characteristics such as production cost, design and production complexity, physical size, etc. multiband performance of the antenna can be argued from different aspects; such that this antenna can eliminate the need to several antennas hence less space will be occupied. Based on the different frequency band of the fifth generation which were explained separately in this chapter, it can be inferred that in the lower band good performance is also required in addition to a small size antenna, while in the upper bands, antenna is small but has a low quantitative and qualitative efficiency, therefore, it requires the use of larger antennas such as horn antennas. Also it should be noted that in these three cases (three frequency bands) MIMO, array and SIW structures are widely used.

Author details

Kioumars Pedram*, Mohsen Karamirad and Negin Pouyanfar

*Address all correspondence to: pedram.qmars@gmail.com

Department of Electrical Engineering, Urmia University, Urmia, Iran

References

[1] Lai HW, Wong H. Substrate integrated magneto-electric dipole antenna for 5G Wi-fi. IEEE Transactions on Antennas and Propagation. 2015;**63**:870-874

[2] Honari MM, Mirzavand R, Melzer J, Mousavi P. A new aperture antenna using substrate integrated waveguide corrugated structures for 5G applications. IEEE Antennas and Wireless Propagation Letters. 2017;**16**:254-257

[3] Ban Y-L, Li C, Wu G, Wong K-L. 4G/5G multiple antennas for future multi-mode smartphone applications. IEEE Access. 2016;**4**:2981-2988

[4] Ali M. Advanced 5G Substrates with Integrated Antennas [thesis]. Atlanta, Georgia: Georgia Institute of Technology; 2017

[5] Saada MHA. Design of Efficient Millimeter Wave Planar Antennas for 5G Communication Systems. Gaza: The Islamic University; 2017

[6] Ojaroudiparchin N, Shen M, Fr G. Multi-layer 5G mobile phone antenna for multi-user MIMO communications. In: 2015 23rd Telecommunications Forum Telfor (TELFOR); 2015. pp. 559-562

[7] Chen Z, Zhang YP. FR4 PCB grid array antenna for millimeter-wave 5G mobile communications. In: 2013 IEEE MTT-S International Microwave Workshop Series on RF and Wireless Technologies for Biomedical and Healthcare Applications (IMWS-BIO); 2013. pp. 1-3

[8] Alreshaid T, Hammi O, Sharawi MS, Sarabandi K. A millimeter wave switched beam planar antenna array. In: 2015 IEEE International Symposium on Antennas and Propagation & USNC/URSI National Radio Science Meeting; 2015. pp. 2117-2118

[9] Yu LC, Kamarudin MR. Investigation of patch phase array antenna orientation at 28GHz for 5G applications. Procedia Computer Science. 2016;**86**:47-50

[10] Ta SX, Choo H, Park I. Broadband printed-dipole antenna and its arrays for 5G applications. IEEE Antennas and Wireless Propagation Letters. 2017;**16**:2183-2186

[11] Asaadi M, Sebak A. High-gain low-profile circularly polarized slotted SIW cavity antenna for MMW applications. IEEE Antennas and Wireless Propagation Letters. 2017; **16**:752-755

[12] Ashraf N, Haraz O, Ashraf MA, Alshebeili S. 28/38-GHz dual-band millimeter wave SIW array antenna with EBG structures for 5G applications. In: 2015 International Conference on Information and Communication Technology Research (ICTRC); 2015. pp. 5-8

[13] Pedram K, Karamirad M, Ranjbaran SMH. A novel circular polarization MIMO antenna in 60 GHz technology. In: 2017 IEEE 4th International Conference on Knowledge-Based Engineering and Innovation (KBEI); 2017. pp. 0335-0338

[14] Jouanlanne C, Clemente A, Huchard M, Keignart J, Barbier C, Le Nadan T, et al. Wideband linearly polarized transmitarray antenna for 60 GHz backhauling. IEEE Transactions on Antennas and Propagation. 2017;**65**:1440-1445

[15] Kaouach H, Dussopt L, Lanteri J, Koleck T, Sauleau R. Wideband low-loss linear and cir-
     cular polarization transmit-arrays in V-band. IEEE Transactions on Antennas and Pro-
     pagation. 2011;59(7):2513-2523

[16] Di Palma L, Clemente A, Dussopt L, Sauleau R, Potier P, Pouliguen P. Circularly polar-
     ized transmitarray with sequential rotation in Ka-band. IEEE Transactions on Antennas
     and Propagation. 2015;63(11):5118-5124

[17] Fixed Radio Systems; Characteristics and Requirements for Point-to-Point Equipment
     and Antennas; Part 4-2: Antennas; Harmonized EN Covering the Essential Requirements
     of Article 3.2 of the R&TTE Directive, document EN 302 217-4-2 V1.5.1. ETSI; 2010

[18] An W, Xu S, Yang F. A two-layer transmitarray antenna. In: Proceedings of the IEEE
     Antennas and Propagation Society International Symposium (APSURSI); Memphis, TN,
     USA. July 2014. pp. 864-865

[19] Clemente L, Dussopt R, Sauleau PP, Pouliguen P. Wideband 400-element electroni-
     cally reconfigurable transmitarray in X band. IEEE Transactions on Antennas and
     Propagation. 2013;61(10):5017-5027

[20] Clemente L, Dussopt R, Sauleau PP, Pouliguen P. Focal distance reduction of transmit-
     array antennas using multiple feeds. IEEE Antennas and Wireless Propagation Letters.
     Nov. 2012;11(11):1311-1314

Where are the Things of the Internet? Precise Time of Arrival Estimation for IoT Positioning

Wen Xu, Armin Dammann and Tobias Laas

Additional information is available at the end of the chapter

http://dx.doi.org/10.5772/intechopen.78063

Abstract

The question how a 5G communication system will look like has been addressed intensely in numerous research projects and in standardization bodies. In the massively connected world of the "Internet of Things" (IoT), it is getting more and more important to be aware of where all these "things" are located. Mobile radio-based technologies envisaged for a 5G system will play an essential role in providing high-accuracy positioning of the "things." In this work, we will first address the fundamental Cramér-Rao lower bound (CRLB) of time of arrival (TOA) estimation in an orthogonal frequency-division multiplexing (OFDM)-based system (such as 4G and 5G) using the pilots. The achievable performance is compared with the 3GPP LTE and potential future 5G requirements. The Ziv-Zakai lower bound (ZZLB) is also considered for TOA estimation, as it is tighter than the CRLB for medium to low signal-to-noise ratios (SNRs). We show how to optimize the waveform in order to reduce the TOA estimation error. Then, we describe some practical low-complexity maximum likelihood (ML) methods for TOA estimation with enhanced first-arriving path detection. Simulation results show that such adaptive ML methods can in some cases (e.g., line of sight) achieve a performance close to the CRLB. Finally, we will briefly discuss cooperation-based positioning, which will become increasingly important for massively connected IoT.

Keywords: Cramér-Rao lower bound, Ziv-Zakai lower bound, time (difference) of arrival, radio-based positioning, cooperative positioning

1. Introduction

Mobile communication has become an integrated part of our daily lives. Today, whereas the state-of-the-art fourth generation (4G) wireless standard *long-term evolution* (LTE) has been in use for a decade, the fifth generation (5G) wireless standard called *new radio* (NR) is being specified for diverse applications in the next 10 years. In the first 5G NR release, Release 15, mainly the *enhanced mobile broadband* (eMBB) use cases have been considered. The *ultra-reliable low latency communication* (URLLC) use cases will be addressed in Release 16. Other use cases such as *massive machine-type communication* (mMTC) as well as the internet of things (IoT) are expected to be taken into account later. Although 5G NR standardization is still underway, a significant amount of details have already been agreed on. One important feature of 4G LTE and 5G NR is the support for accurate positioning of a user equipment (UE), i.e., the estimation of the position of the UE or the "thing" in the network, such as a car, a drone, etc. Especially in the massively connected world of IoT, it is getting more and more important to be aware of where all these things are located. Mobile radio-based technologies envisaged for a 5G system will play an essential role in providing high-accuracy positioning of the "things."

2. Overview on mobile radio positioning techniques

The 2G, 3G, and 4G cellular communication standards have specified a variety of positioning methods. These methods infer position information from received signals and include Cell-ID, received signal strength (RSS) as well as time difference of arrival (TDOA)-based methods. All these methods have in common that they use downlink signals. Propagation delay-based methods like TDOA require signal reception from three base stations (BSs) in order to calculate a 2D UE position as shown in **Figure 1**. To estimate the position in 3D, at least four BSs are needed. In many environments, the probability of receiving signals from three different BSs with sufficient quality has shown to be quite low. In the example shown in **Figure 1**, it is not possible to get a position fix for UE$_3$ since it receives the signal from BS$_3$ only. For increasing adjacent BS hearability, the idle period downlink (IPDL) has been implemented in 3G UMTS [1]. LTE has

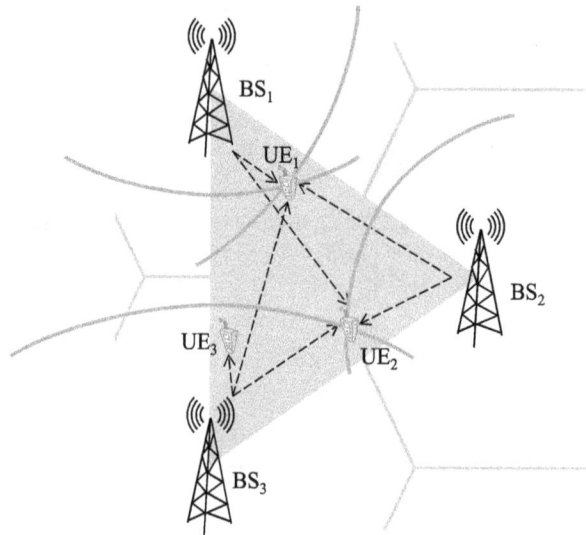

Figure 1. Today's cellular mobile system, where UEs require signals from at least three different BSs in order to calculate their position in 2D. The UEs operate independently from each other without any cooperation.

addressed this problem since its Release 9 with the specification of positioning reference signals (PRSs) [2]. However, multipath and non-line-of-sight (NLOS) propagation are still present and potentially cause severe positioning performance degradations. Usually, the probability of receiving signals under line-of-sight (LOS) condition decreases with increasing distance between BS and UE [3].

For 4G LTE, UE positioning is defined in [4]. There are two protocols, the LTE Positioning Protocol (LPP) [5], which specifies the protocol between the UE and the so-called location server, and LTE Positioning Protocol Annex (LPPa) [6], which specifies the protocol between the BS and the location server. There are the following methods in Release 13 [4]: (1) Observed TDOA (OTDOA), (2) assisted-global navigation satellite system (A-GNSS), (3) enhanced cell-ID (E-CID), (4) barometric sensor, (5) terrestrial beacon system (TBS), (6) WLAN, and (7) Bluetooth. The first three techniques have been in the standard since Release 9, and the next four have been added in Release 13 to fulfill the new FCC wireless indoor E-911 location accuracy requirements from 2015 [7]. A-GNSS is used to provide assistance data to the GNSS receiver in the UE. E-CID is a coarse positioning method, which can use the Cell-ID, the received signal power/quality at the UE, timing information, and the angles of arrival (AOA) at the BS to estimate the position. Barometric sensor positioning uses a barometric sensor to identify the height of the UE. WLAN positioning can use the (B)SSID of WLAN access points near the UE together with the RSS indicator (RSSI) and round-trip time. Bluetooth positioning can use Bluetooth beacon identifiers near the UE together with the RSSI. TBS can use Metropolitan Beacon Systems (MBS), a network of ground-based transmitters broadcasting high-precision time signals similar to the global positioning system (GPS).

LTE Release 11 also adds support for uplink TDOA, which means that the UE sends pilots for positioning and several BSs to measure the TDOA. LTE Release 13 and, recently 14, addressed positioning for "further enhancements for enhanced machine-type communication" (feMTC) and NarrowBand IoT (NB-IoT).

We focus here on downlink OTDOA, which is a multilateration method, as shown in **Figure 1**. Several BSs send the PRS to one UE, which estimates the TDOA with respect to some reference BS, and feeds back the (quantized) TDOA to the location server. Each TDOA measurement restricts the location of the UE to a hyperboloid. The location server then estimates the position of the UE based on the TDOAs.

The LTE standard specifies a set of downlink pilots or reference signals (RSs) with different time-frequency patterns, such as those shown in **Figure 3**. Note that the BS does not transmit on the data channels in the resource blocks used for the PRS. There are six possible frequency shifts for BSs operating at the same frequency. The PRS is repeated periodically. In order to further increase precision, the PRS of certain BSs can be muted in certain repetitions to reduce interference. The PRS of other BSs can be sent on the same or in a different frequency band as the serving BS. An overview can be found in [8].

2.1. UE positioning requirements

Services and applications based on accurate knowledge of the user position, such as location-sensitive billing, fraud detection, fleet management, and intelligent transportation systems have become increasingly important. In 1996, the United States Federal Communications Commission (FCC) mandated all US wireless network operators and mobile devices to provide location information for Enhanced-911 (E-911) [9]: caller location must be provided to public-safety answering points (PSAPs) with 50 m accuracy for 67% of calls and 150 m accuracy for 95% of calls. In 2015, the FCC published the wireless indoor E-911 location accuracy requirements [7]. They include but are not limited to that within 6 years; for 80% of all wireless 911 calls, the horizontal location of the caller must be known within 50 m and the vertical location must fulfill some z-axis metric that still has to be approved by the FCC. Alternatively, the so-called dispatchable location can be provided, which is the address of the building together with a floor or apartment number. Furthermore, barometric sensor data must be made available for all UEs that support it.

The FCC requirements can be met by GNSS such as GPS in many environments. Typically, the GPS for civil applications can provide a positioning accuracy of a few meters. However in some cases, such as indoors or in urban canyons, the GPS signal may be too weak or scattered too much to provide the required accuracy. As a complement, wireless systems like GSM, UMTS, or LTE provide good coverage in such scenarios. Accordingly, requirements for TOA and TDOA measurements have been specified in 3GPP LTE Release 9 to ensure accurate UE positioning even under bad conditions (e.g., with channels quickly varying and SNRs being as low as -13 dB). Depending on use cases, 5G will have much stricter requirements; e.g., for V2X vulnerable road user discovery, accuracy as high as 10 cm may be required (see [10]).

2.2. What can 5G new radio (NR) do better for positioning?

3GPP, which is responsible for 5G standardization, has decided that OFDM will be used for 5G NR, as in 4G LTE. Specifically, the following parameters have been agreed on (see [11]):

1. *Subcarrier spacing (SCS):* for 4G, the subcarrier spacing is fixed to 15 kHz, except for multicast-broadcast single-frequency network (MBSFN) services for which a subcarrier spacing of 7.5 kHz is used. In contrast, 5G will deploy multiple subcarrier spacings ranging from 15 to 480 kHz, which are all integer multiples of 15 kHz.

2. *Cyclic prefix (CP):* 5G has adopted the same approach as 4G, where the CP can be either normal CP (NCP) or extended CP (ECP). The choice of the CP depends on the expected signal dispersion. In 5G, ECP is expected to be associated with the 60 kHz subcarrier spacing.

3. *Frame structure:* in 4G, the transmission time interval (TTI) was specified to be 1 ms, which is the subframe duration, and a subframe consists of two slots. In 5G, a subframe can contain 1, 2, 4, 8, 16, and 32 slots. For both 4G and 5G, each slot consists of 14 OFDM symbols.

4. *Bandwidth:* a single carrier of 5G is expected to support a bandwidth of up to 100 MHz for carrier frequency below 6 GHz, and up to 400 MHz for high (millimeter wave) carrier frequency. This leads to much higher accuracy for radio-based positioning.

Until now, 5G NR positioning has not yet been specified. Hence, this work will take 4G as an example. As 4G and 5G both employ OFDM and similar frame structure, the results obtained for

4G can be reasonably extrapolated to 5G; e.g., with a similar pilot signal density for TOA estimation, when 5G has five times the bandwidth of 4G, it can achieve five times as high positioning accuracy. Also, the following features envisaged in 5G are beneficial for positioning.

- *Higher frequencies and large signal bandwidths*: larger signal bandwidths allow a better resolution of the wireless channel in time, and therefore, more accurate estimation of multipath components, in particular, their signal propagation delays. In addition to the conventional frequency bands from about 450 MHz to 6 GHz ([12], Section 5.5), 5G will also use millimeter wave frequency bands, e.g., at 28 or 60 GHz. At those frequencies, the attenuation of the channel is high, since the antennas need to be smaller for a similar directivity. This increases the probability of LOS reception conditions as any NLOS condition is likely to be blocked and reduces the risk of positioning errors due to the NLOS bias. Furthermore, higher frequencies together with massive multiple-input-multiple-output (MIMO) schemes allow tracking the individual terminals by beam forming with antenna arrays more accurately (see e.g., [10]).

- *Dense networks*: a denser grid of BSs reduces distances between UEs and BSs. With lower BS-UE distances, the probability of LOS signal reception increases. This reduces the risk of positioning errors due to the NLOS bias.

- *Device-to-device (D2D) communications with a large number of connected devices*: additional links provide additional signal observations that can be exploited to determine pseudoranges among UEs as shown in **Figure 2**. With D2D communication capabilities, UEs are inherently receiving signals from each other. Signal processing entities for D2D communications, in particular synchronization and channel estimation units, can be reused for signal propagation delay estimation. D2D communication provides a meshed

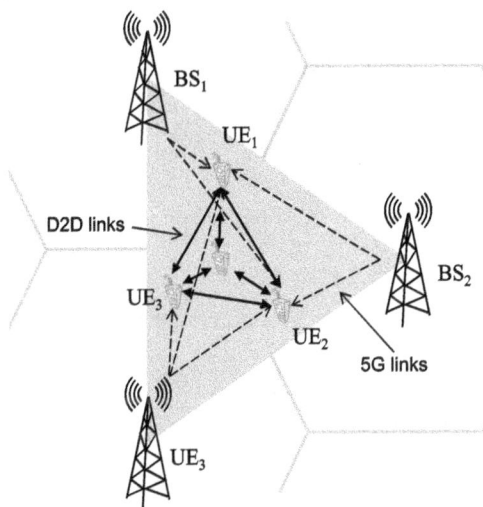

Figure 2. 5G envisages D2D communications, where UEs may cooperate with each other for positioning. If the mesh of D2D links is sufficiently dense, positioning works even if there are less than three BSs visible to individual UEs.

network structure rather than the star-shaped one for today's mobile cellular systems. Assuming a fully connected mesh as a best case, the number of D2D links grows quadratically with the number of UEs N_{UE}. As the number of unknown positions increases linearly with N_{UE}, D2D links provide significant redundancy in the number of observations to neglect links under disadvantageous propagation conditions like low SNR, NLOS, severe multipath, bad geometry, etc. Even unknowns like NLOS bias terms can be estimated with a sufficient number of observations. Consequently, precise positioning can be achieved by exploiting cooperation among UEs.

3. Cramér-Rao and Ziv-Zakai lower bound in an OFDM system

In this section, we will describe the Cramér-Rao lower bound (CRLB) and Ziv-Zakai lower bound (ZZLB) for T(D)OA estimation in an OFDM system transmitting over an additive white Gaussian noise (AWGN) channel. The CRLB follows the derivation in [13], but additionally allows a frequency shift of the subcarriers, which is needed for NB-IoT. Consider the following OFDM transmit signal (without the CP)

$$s_l[n] = \frac{1}{\sqrt{N}} \sum_{k=-N/2}^{N/2-1} S_l[k] \exp\left(j\frac{2\pi}{N}(k+\kappa)n\right), \quad 0 \leq \kappa < 1, \tag{1}$$

where $S_l[k]$ is the signal allocated to the kth subcarrier of the lth OFDM symbol, N is the number of subcarriers and κ shifts the subcarriers in frequency domain. Let us transform this signal into continuous time domain to estimate the continuous delay τ, the TOA. By removing the periodic replicas in frequency domain by multiplying with the rectangular function

$$\text{rect}(\omega) = \begin{cases} 1 & \text{for } \frac{-\pi}{T} \leq \omega < \frac{\pi}{T}, \\ 0 & \text{else}, \end{cases} \tag{2}$$

we have the frequency-domain representation

$$S_l(\omega) = \sqrt{\frac{2\pi}{N}} \sum_{k=-N/2}^{N/2-1} S_l[k]\delta\left(\omega - (k+\kappa)\frac{2\pi}{NT}\right), \tag{3}$$

where T is the sampling time interval. In time domain, the same signal becomes

$$s_l(t) = \frac{1}{\sqrt{N}} \sum_{k=-N/2}^{N/2-1} S_l[k] \exp\left(j2\pi(k+\kappa)\frac{t}{NT}\right). \tag{4}$$

Then, we sample this signal delayed by τ for $n = 0, ..., N-1$,

$$s_{R,l}[n] := s_l(nT - \tau) = \frac{1}{\sqrt{N}} \sum_{k=-N/2}^{N/2-1} S_l[k] \exp\left(j2\pi(k+\kappa)\Delta f(nT - \tau)\right), \quad \Delta f = \frac{1}{NT}. \tag{5}$$

Now consider the system transmitting over an AWGN channel

$$y_l[n] = s_{R,l}[n] + z_l[n], \quad z_l[n] \sim \mathcal{CN}(0, \sigma^2). \tag{6}$$

The variance and the CRLB for any unbiased estimate $\hat{\tau}$ of τ from the measurement vector $\mathbf{y} = [y[0], \ldots, y[N-1]]^T$ become ([14], Chapter 3)

$$\mathrm{Var}(\hat{\tau}(\mathbf{y})) \geq \mathrm{CRLB}(\hat{\tau}) = \frac{1}{\mathrm{E}\left[\left(\frac{\partial}{\partial \tau} \ln p(\mathbf{y}|\tau)\right)^2\right]} \tag{7}$$

as long as the regularity condition $\mathrm{E}\left[\frac{\partial \ln p(\mathbf{y}|\tau)}{\partial \tau}\right] = 0$ $\forall \tau$ is fulfilled. For the AWGN channel, the CRLB can be expressed as

$$\mathrm{CRLB}(\hat{\tau}) = \frac{\sigma^2}{2 \sum\limits_{l=0}^{N_{\text{symb}}-1} \sum\limits_{n=0}^{N-1} \left|\frac{\partial}{\partial \tau} s_{R,l}[n]\right|^2}. \tag{8}$$

when N_{symb} OFDM symbols are used to estimate τ.

This expression can be simplified as

$$\sum_{n=0}^{N-1} \left|\frac{\partial}{\partial \tau} s_{R,l}[n]\right|^2 = \sum_{n=0}^{N-1} \left|\frac{\partial}{\partial \tau} \frac{1}{\sqrt{N}} \sum_{k=-N/2}^{N/2-1} S_l[k] \exp\left(j2\pi(k+\kappa)\Delta f(nT - \tau)\right)\right|^2$$

$$= \frac{4\pi^2(\Delta f)^2}{N} \sum_{m=-N/2}^{N/2-1} \sum_{k=-N/2}^{N/2-1} (m+\kappa)(k+\kappa) S_l^*[m] S_l[k] \exp\left(j\frac{2\pi}{N}(k-m)n\right) N\delta_{mk}, \tag{9}$$

where the Kronecker delta δ_{mk} comes from the orthogonality of the subcarriers, i.e.,

$$\sum_{n=0}^{N-1} \exp\left(j2\pi\Delta f(m-k)nT\right) = N\delta_{mk}. \tag{10}$$

In this way, we obtain the CRLB [13]

$$\mathrm{Var}\{\hat{\tau}\} \geq \mathrm{CRLB}(\hat{\tau}) = \frac{\sigma^2}{8\pi^2(\Delta f)^2 \sum\limits_{l=0}^{N_{\text{symb}}-1} \sum\limits_{k=-N/2}^{N/2-1} (k+\kappa)^2 S_l[k]^2}. \tag{11}$$

As the PRS does not carry any time stamp, there are ambiguities in the TOA estimation. Therefore, TDOA estimation is usually used. The CRLB for TDOA estimation is

$$\mathrm{CRLB}_{\mathrm{TD}} = \mathrm{CRLB}(\hat{\tau}_{\text{ref}}) + \mathrm{CRLB}(\hat{\tau}), \tag{12}$$

where τ_{ref} is the time delay to the reference BS, which, e.g., serves the UE, $\hat{\tau}_{\text{ref}}$ is its estimate, and we assume that τ and τ_{ref} are statistically independent. We will focus on $\text{CRLB}(\hat{\tau})$ in the following and abbreviate its standard deviation as

$$\sigma_{\text{CRLB}} = \sqrt{\text{CRLB}(\hat{\tau})}. \tag{13}$$

Note that this is the standard deviation of the TOA, which can be translated to a standard deviation of the distance d by multiplying with the speed of light in free space c_0, i.e.,

$$\text{CRLB}\left(\hat{d}\right) = c_0\text{CRLB}(\hat{\tau}), \tag{14}$$

where $\hat{d}$ is the estimate of d.

3.1. Achievable TOA measurement accuracy using pilots in 3GPP LTE

Different pilots or RSs have been specified in LTE, e.g., the primary synchronization signal (PSS), the secondary synchronization signal (SSS), and the cell-specific RS (CRS). In general, all or parts of these pilots can also be used for TDOA estimation (see **Figure 3**). The graphs in **Figure 4** show σ_{CRLB}, computed according to Eqs. (11) and (13), using different pilots specified in LTE and one receive antenna, with $E_s := E\left\{|S_l(k)|^2\right\}$ being constant for the pilots specified in LTE systems, such as the PSS, SSS, CRS, and PRS. Here, a subframe of 1 ms contains 14 consecutive OFDM symbols, as in the case of the LTE normal CP. The PSS, SSS, CRS, and PRS are mapped to the corresponding resource elements [2]. Note the lowest bound (i.e., the highest measurement accuracy) is obtained by utilizing *all* the four pilots (PSS, SSS, CRS, and PRS) simultaneously. Among all available pilots in LTE, the PRS, as expected, achieves the highest accuracy in terms of the CRLB since it almost spans the whole bandwidth and there are also more PRS symbols available than, say, CRS symbols (see **Figure 3**). As it can be seen, using the PRS instead of the CRS can have a gain of about 3 dB. When CRS in addition to PRS is used, about 1 dB can be gained. As shown in the next section, an adaptive ML detector can have an estimation accuracy close to the CRLB, especially for scenarios where the first path is dominant.

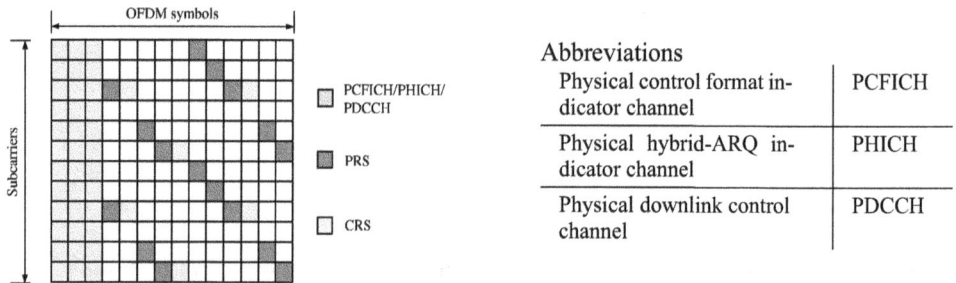

	Abbreviations	
☐ PCFICH/PHICH/ PDCCH	Physical control format in- dicator channel	PCFICH
■ PRS	Physical hybrid-ARQ in- dicator channel	PHICH
☐ CRS	Physical downlink control channel	PDCCH

Figure 3. An example LTE signal pattern with a cell-specific reference signal (CRS) and positioning reference signal (PRS) in a single physical resource block (PRB).

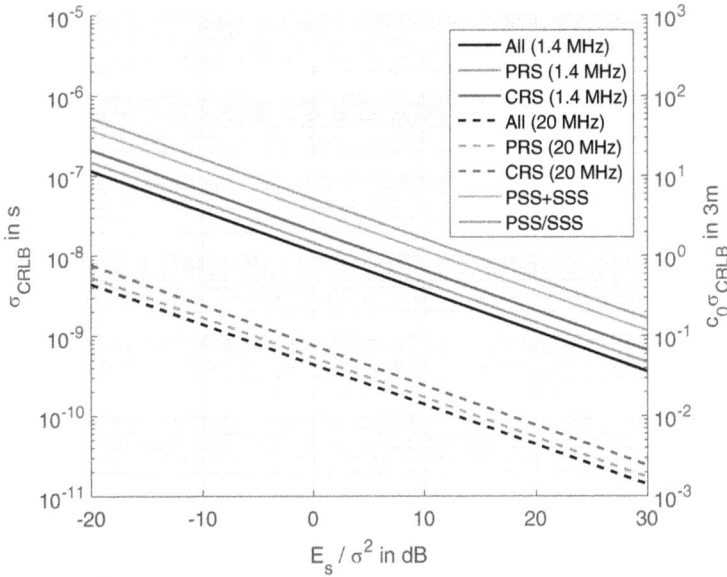

Figure 4. CRLB for TOA measurement using different pilots in one subframe.

3.2. CRLB for LTE and LTE NB-IoT

In what follows, we will focus on the PRS. LTE supports several bandwidths of the PRS, from 6 to 100 physical resource blocks (PRBs), consisting of 12 subcarriers each. The PRBs are placed symmetrically around the carrier frequency. Then, the usable bandwidth in the downlink is 1.095 to 18.015 MHz, where the additional 15 MHz come from the additional unused DC subcarrier. $\kappa = 0$ corresponds to the conventional LTE downlink without considering the DC subcarrier. For bandwidths between 1.4 and 20 MHz [12], the (nominal) sampling rate of the system is $T = 16T_s$ to T_s, where $T_s = 1/30.72\,\mu s \approx 32.552\,ns$ is the *LTE basic time unit*. The (nominal) FFT size changes accordingly from $N = 128$ to 2048 (c.f. **Table 1**).

Consider the PRS for a normal CP when there are only one or two physical broadcast channel (PBCH) antenna ports ([2], Chapter 6.10). In each subframe for positioning, the PRS occupies 8 out of 14 OFDM symbols, as shown in **Figure 3**. In each PRB, two subcarriers are allocated for the PRS in each of the eight OFDM symbols. If we average over the six different cyclic shifts, the PRS corresponds to an equal power allocation over all usable subcarriers. **Figure 5** shows the CRLB for different bandwidths of the PRS for a single subframe having the same subframe sum power allocated to the PRS

$$P_{sum} := \sum_{l=0}^{N_{symb}-1} \sum_{k=-N/2}^{N/2-1} |S_l[k]|^2. \tag{15}$$

In 5G, for $\Delta f = 15\,kHz$, the supported range of channel bandwidths is from 5 to 50 MHz (c.f. **Table 1**). In general, the usable downlink bandwidth is larger for the same channel bandwidth,

Channel bandwidth	Number of PRBs	Usable downlink bandwidth	Nominal FFT size	T
NB-IoT: 200 kHz	1	180 MHz	128	$16T_s$
LTE: 1.4 MHz	6	1.095 MHz	128	$16T_s$
LTE: 3 MHz	15	2.715 MHz	256	$8T_s$
LTE: 5 MHz	25	4.515 MHz	512	$4T_s$
LTE: 10 MHz	50	9.015 MHz	1024	$2T_s$
LTE: 15 MHz	75	13.515 MHz	1536	$4/3T_s$
LTE: 20 MHz	100	18.015 MHz	2048	T_s
5G: 5	25	4.5 MHz	512	$4T_s$
5G: 10	52	9.36 MHz	1024	$2T_s$
5G: 15	79	14.22 MHz	1536	$4/3T_s$
5G: 20	108	19.44 MHz	2048	T_s
5G: 25	133	23.94 MHz	3072	$2/3T_s$
5G: 40	216	38.88 MHz	4096	$T_s/2$
5G: 50	270	40.5 MHz	4096	$T_s/2$

Table 1. Some 4G LTE parameters [12] and 5G NR parameters for $\Delta f = 15\,\mathrm{kHz}$ [11].

as more physical resource blocks are used. No PRS has been standardized for 5G yet. To estimate the future accuracy of 5G, we assume the LTE PRS but extended to the different number of PRBs. As we can see in **Figure 5**, the accuracy of the same channel bandwidth is expected to be slightly better than in LTE since the usable downlink bandwidth increases.

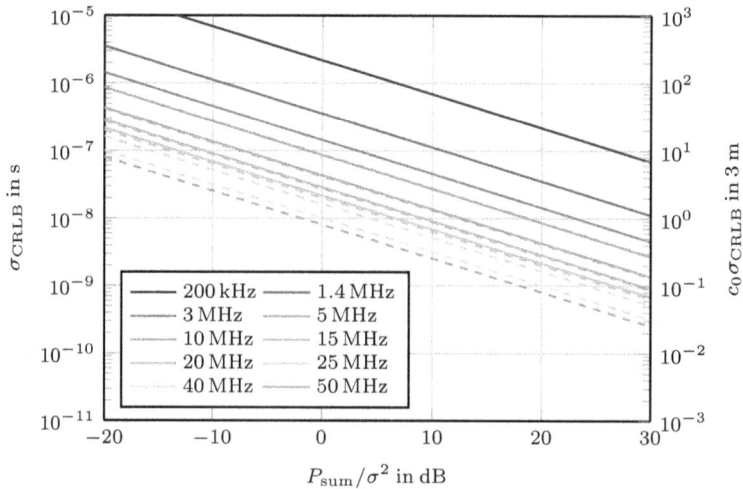

Figure 5. CRLB of LTE (N)PRS with different bandwidths but with the same sum power for a single subframe (solid), including 5G performance estimate using the LTE PRS adjusted to the different number of physical resource blocks (dashed).

Now let us consider IoT. Accurate positioning for IoT is challenging due to the small channel bandwidth for machine type communication (1.4 MHz in LTE-M) and NB-IoT systems (200 kHz in LTE NB-IoT). As $\sigma_{CRLB} = \mathcal{O}\left((\Delta f N)^{-1}\right)$, the positioning accuracy reduces considerably compared to a 10 MHz PRS with the same energy (c.f. **Figures 5** and **6**). Therefore, the PRS has been optimized for LTE-M, and a new narrowband PRS (NPRS) has been introduced for NB-IoT in Release 14 [15]. Those improvements decrease the so-called periodicity of the PRS or increase the length of the PRS, up to about 0.5 s. This increases the energy spent on the PRS and decreases the efficiency of the system meaning that fewer resources can be allocated to data transmission. In order to increase the effective bandwidth of the PRS, which decreases the energy and time needed for the PRS, LTE-M supports frequency hopping, but NB-IoT currently only supports artificial frequency hopping by configuring the PRS onto multiple NB-IoT carriers [15], see e.g., [16] for a study on its performance. Frequency hopping in NB-IoT can be more difficult than in LTE-M since NB-IoT can operate in individual small unused gaps in the spectrum, while LTE-M uses (parts of an) LTE channel. Similarly, in order to increase the effective bandwidth, carrier aggregation can be employed [17].

In contrast to the conventional LTE downlink, we have $\kappa = 1/2$ for NB-IoT—at least in guard band and standalone operation mode. NB-IoT occupies one PRB, i.e., the 12 subcarriers $-6, ..., 5$ in the downlink [2]. The NPRS occupies at least 10 subframes consisting of 14 subsequent OFDM symbols each, which are all used for the NPRS in guard band and standalone operation mode [2]. Depending on the cell ID, the NPRS is shifted circularly in the occupied subcarriers by $v \in \{0, ..., 5\}$ subcarriers. In each subframe, each subcarrier is allocated twice with the NPRS, except for subcarriers $k = v$ and $k = v - 6$, which are allocated four times. The allocation pattern is similar to the one for the PRS (c.f. **Figure 3**). Let P_{symb} denote the power of each allocated symbol in a subcarrier. Then for one subframe,

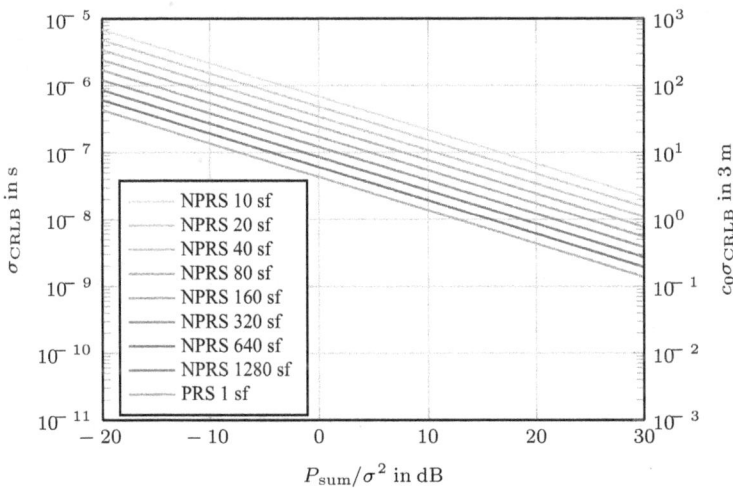

Figure 6. CRLB of the LTE NPRS with different numbers of subframes and of the LTE PRS for a 10-MHz channel.

$$\sum_{l=0}^{13} \sum_{k=-N/2}^{N/2-1} (k+\kappa)^2 |S_l[k]|^2 = \frac{1}{28}\left(4\nu^2 - 20\nu + 347\right)P_{\text{sum}}, \quad P_{\text{sum}} = 28P_{\text{symb}} \tag{16}$$

holds, where the nominal FFT size for NB-IoT is $N = 128$ (c.f. **Table 1**). That means simplifying the average CRLB for TOA estimation yields

$$\text{CRLB}_{\text{NPRS, avg}}(\hat{\tau}) = \frac{28\sigma^2}{8\pi^2(\Delta f)^2(333 + 2/3)P_{\text{sum}}}. \tag{17}$$

As for the LTE PRS, this corresponds to an equal power allocation on all usable subcarriers, but with $\kappa = 1/2$ instead of $\kappa = 0$. As shown in **Figure 6**, the positioning accuracy of the NPRS, even with 1280 subframes, is still a bit worse than the 10 MHz PRS with one subframe.

3.3. ZZLB and waveform optimization

In LTE, all PRS and NPRS symbols are sent with the same power. But when we consider the CRLB in Eq. (11), the optimum power allocation strategy is to allocate all power to positioning symbols in those subcarriers that are furthest from the center frequency. That means, compared to Eq. (17), we get for one subframe

$$\sum_{l=0}^{13} \sum_{k=-N/2}^{N/2-1} (k+\kappa)^2 |S_l[k]|^2 = 14 \cdot 2 \cdot 5.5^2 P_{\text{symb}} = 847P_{\text{symb}} = \frac{121}{4}P_{\text{sum}}. \tag{18}$$

So σ_{CRLB} improves by a factor of about 1.6. Consequently, the TOA estimation time with the NPRS can be reduced from about 0.5 to about 0.31 s, in order to achieve the same accuracy.

In practice, however, when we approximate the power allocation by a Dirac in the edge frequencies, this waveform is not optimal for all SNRs in general—especially at low SNR—since its autocorrelation has got large sidelobes, which the estimator can confuse with the main lobe [18] (see **Figure 7**). There are tighter bounds for the estimation error that take this into account, e.g., the ZZLB [19, 20], which is given by

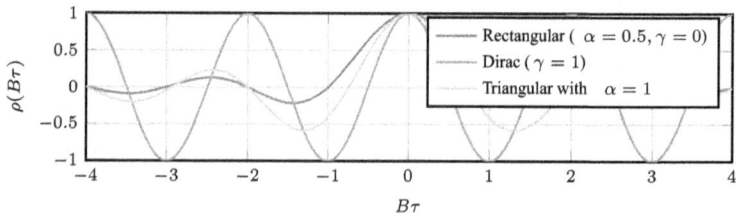

Figure 7. Comparison of the normalized autocorrelation of the Dirac-rectangular and the triangular waveform.

$$\text{ZZLB} = \int_0^{T_a} \tau\left(1 - \frac{\tau}{T_a}\right)\Phi\left(\sqrt{\frac{P_{\text{sum}}}{\sigma^2}(1 - \rho(\tau))}\right)d\tau, \quad \Phi(x) = \frac{1}{\sqrt{2\pi}}\int_x^{+\infty} e^{-u^2/2}du, \quad (19)$$

for TOA estimation, where we assume the prior information that τ is uniformly distributed in $[0, T_a]$ with the observation time interval T_a and where $\rho(\tau)$ is the real-valued normalized autocorrelation function of the positioning symbols. In LTE, the UE gets the required prior information on the expected delay via the LPP [5]. Note that the optimum waveform w.r.t. ZZLB also depends on the SNR, in contrast to the CRLB. Therefore, depending on the region of the SNR of interest, the waveform that minimizes the ZZLB may be different.

In [18], the optimization w.r.t. ZZLB in continuous time over a so-called triangular waveform with the parameter $\alpha \in [0, 1]$ and over a Dirac-rectangular waveform with parameter $\gamma \in [0, 1]$ with power spectral densities

$$|S_{\text{tri}}(f)|^2 = \begin{cases} (1 - \alpha)\frac{2}{B} - \frac{4(1 - 2\alpha)}{B^2}|f|, & |f| \le \frac{B}{2}, \\ 0 & |f| > \frac{B}{2}, \end{cases} \quad (20)$$

$$|S_{\text{dr}}(f)|^2 = \begin{cases} \frac{1-\gamma}{B} + \frac{\gamma}{2}\left[\delta\left(f + \frac{B}{2}\right) + \delta\left(f - \frac{B}{2}\right)\right], & |f| \le \frac{B}{2}, \\ 0, & |f| > \frac{B}{2} \end{cases} \quad (21)$$

is shown, where B is the bandwidth (see **Figure 8**). **Figure 9** shows the ZZLB, with.

$$\sigma_{\text{ZZLB}} := \sqrt{\text{ZZLB}}. \quad (22)$$

For a very low SNR, the σ_{ZZLB}s are slightly below $T_a/\sqrt{12}$, which is the standard deviation of the uniform distribution we use as a priori information for the ZZLB. The CRLB does not consider the a priori information. For a high SNR, the ZZLB converges to the CRLB. For NB-IoT, we have $B \approx 11\Delta f = 165$ kHz and thus $T_a \approx 60.6\,\mu$s for the same configuration as in [18]. There it was shown that the optimal triangular waveform is the one with $\alpha = 1$, but for the Dirac-rectangular waveform, the optimum value of γ depends on P_{sum}/σ^2. For small P_{sum}/σ^2, a

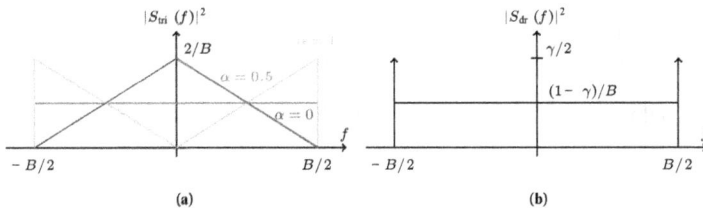

Figure 8. Power spectral density of the triangular and the Dirac-rectangular waveform (a) Triangular, (b) Dirac-rectangular.

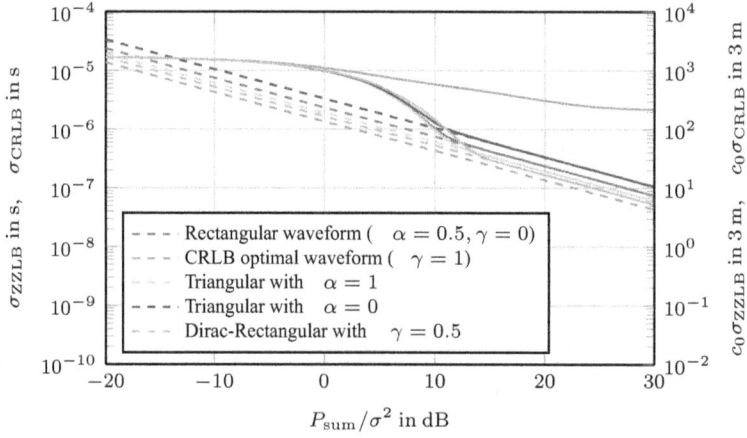

Figure 9. ZZLB of the Dirac-rectangular waveform and the triangular waveform for different parameters α, γ and the unrealistically large $T_a \approx 60.6\,\mu s$.

triangular waveform is better than a Dirac-rectangular waveform, but for large P_{sum}/σ^2, it is the other way round. We observe the same behavior in **Figure 9**.

Sampling the Dirac-rectangular waveform with $\gamma < 1$ or the triangular waveform with $\alpha = 1$ could be a good candidate for the power allocation of the PRS for IoT in 5G, but the positioning accuracy will not improve by the full factor of 1.6, corresponding to the CRLB optimal allocation.

4. Practical TOA estimation based on first tap detection

4.1. Maximum likelihood (ML) timing estimation

The reference signal $s_l(n)$, such as CRS and/or PRS, is embedded in the received signal $y_l(n)$. The target of the TOA estimation is to determine the position τ in the received signal y_l, say, using the maximum likelihood (ML) criterion. Notice that the ML estimator has the asymptotic properties of being unbiased and achieving the CRLB [14]. Consider the other paths as interference, the ML criterion for timing estimation of the first path reduces to a correlation-based criterion. The correlation-based method can be realized in time or frequency domain. In the following, we focus on the time domain-based method [13].

The received signal $y_l(n)$ is correlated with the replica of the transmitted signal $s_l(n)$, i.e.,

$$R(t) := \sum_{l=0}^{N_{symb}-1} \sum_{n=0}^{N-1} y_l(n+t)s_l^*(n), \qquad t \in [0, W-1], \tag{23}$$

where $W = 2G$ is chosen as the search window size. To ease the analysis, we first assume $s_l(n)$ has ideal autocorrelation property, and the power of the transmit signal $s_l(n)$ is P_s. Then, with some derivations, the correlation can be written as

$$R(t) = P_s \sum_{l=0}^{N_{symb}-1} h_l(t - \tau) + R_{res}(t), \tag{24}$$

where $R_{res}(t)$ represents the total residual noise and interference part resulting from correlation between $s_l(n)$ and $y_l(n + t)$.

Assume the channel is unknown but remains invariant for N_{symb} OFDM symbols, $h_l(t) = h(t)$, then the noncoherent detector can be employed. The metric for the TOA detection, which is also called the *correlation profile*, is given here by

$$\Lambda(t) := E\left\{|R(t)|^2\right\} = N_{symb}^2 P_s^2 \gamma(t - \tau) + P_s N_{symb}\sigma^2, \tag{25}$$

with $\gamma(t) := E\left\{|h(t)|^2\right\}$. $E\left\{|R(t)|^2\right\}$ is used to denote the statistical average of $|R(t)|^2$ over multiple subframes containing RS signals. In LTE, a group of several consecutive subframes containing the RS is sometimes referred to as a positioning occasion. Usually, an LTE positioning measurement is done on one or more occasions.

4.2. Signal arrival region determination

We now determine the arrival region of the RS. For a multipath channel, the signal arrival region will have multiple taps corresponding to the taps of the channel. The signal paths are, therefore, reflected by the channel paths. A moving sum is computed as

$$\Lambda_{win}(u) = \sum_{t=u}^{u+G-1} E\left\{|R(t)|^2\right\}, \qquad u \in [0, G-1] \tag{26}$$

The signal can be regarded as arrived in the time region

$$u_0 \leqslant t \leqslant u_0 + G - 1 \quad \text{s.t.} \quad u_0 = \arg\max_u \{\Lambda_{win}(u)\} \tag{27}$$

When $t - \tau \geqslant L$ or $t - \tau < 0$, only the noise power related term, the noise floor $N_f := P_s N_{symb}\sigma^2$ remains in the correlation. N_f, which is used here as σ^2, can be calculated by averaging the terms outside the signal region.

For a single path channel, such as in the case of the LOS signal, the TOA can be detected, by searching for the path with the strongest signal power. For a multipath channel, in particular, when the first arriving path is not the strongest (e.g., under ETU channel), the TOA estimation becomes biased. Usually, a threshold is needed to determine the first arriving path. Especially in the case of strong noise and interference from multiple cells, the metric $\Lambda(t)$ may not provide sufficient accuracy. Consider

$$\frac{E\{|R(t)|^2\}}{N_f} = \frac{N_{\text{symb}}P_s\gamma(t-\tau)}{\sigma^2} + 1 = N_{\text{symb}}SNR(t) + 1, \tag{28}$$

where $SNR(t) := \frac{P_s\gamma(t-\tau)}{\sigma^2}$ is the SNR for each correlation sample, and it holds

$$SNR(t) = \frac{1}{N_{\text{symb}}}\left(\frac{E\{|R(t)|^2\}}{N_f} - 1\right). \tag{29}$$

The following criterion then takes a fixed SNR value as a threshold to estimate the first tap

$$\tau = \min_{u_0 \leqslant t < u_0+G-1} t \quad \text{s.t.} \quad SNR(t) \geqslant SNR_{\text{th}}. \tag{30}$$

SNR_{th} is the required SNR for detection, which can be set as, e.g., -13 dB for 3GPP Release 9 OTDOA measurement.

4.3. Adaptive-threshold-based first tap detection

As $s_l(n)$ is not ideally autocorrelated, the noise floor N_f would contain further terms besides $P_s N_{\text{symb}}\sigma^2$. Therefore, we can express the noise floor as

$$N_f = P_s N_{\text{symb}}\sigma^2 + \varepsilon(N_{\text{symb}}), \tag{31}$$

where $\varepsilon(N_{\text{symb}})$ is a parameter related to N_{symb}, $s_l(n)$, and $y_l(n)$ and $y_l(n)$ is in turn dependent on the channel and the interference.

Here, we use a criterion, which jointly considers the noise power and the received *signal* power to determine a varying (adaptive) detection threshold. Assume the metric peak relying on the signal power and noise is

$$\Lambda_{\max} = \max_{u_0 \leqslant t < u_0+G-1} E\{|R(t)|^2\}. \tag{32}$$

The adaptive threshold can then be defined as $\Lambda_{\text{th}} = \alpha\sqrt{\Lambda_{\max}N_f}$. Alternatively, the threshold can be defined as $\Lambda_{\text{th}} = \alpha(\beta\Lambda_{\max} + (1-\beta)N_f)$, where α is a design parameter, $\beta \in [0,1]$ is a constant trading-off between the noise floor and the metric peak. α and β were determined through simulations, to have a trade-off for different channels and different SNRs.

Given the threshold, the criterion for the adaptive threshold detection can be expressed as

$$\tau = \min_{u_0 \leqslant t < u_0+G-1} t \quad \text{s.t.} \quad E\{|R(t)|^2\} \geqslant \Lambda_{\text{th}}. \tag{33}$$

This criterion usually leads to better performance especially under a multipath channel, and in some cases can achieve performance close to the CRLB [13].

LTE supports bandwidths up to 20 MHz, which corresponds to a sampling rate of 30.72×10^6 samples/s in the baseband signal. For a bandwidth smaller than 20 MHz, the processing can be done at a smaller sampling rate to reduce the processing load; e.g., when the channel has a bandwidth of 1.4 MHz, 1/16 of the rate is sufficient. However, better accuracy can be obtained by a higher sampling rate due to the receive diversity gain.

5. Cooperation for accurate and reliable mobile radio positioning

5.1. Cooperative positioning principle

Future wireless technologies such as 5G enable UEs to cooperate with each other. By mutually observing their transmitted signals, UEs can estimate the ranges among themselves. If the mesh of mutually observed D2D links is sufficiently dense, positioning works even if there are less than three BSs visible to individual UEs as shown in **Figure 2**. For ranging, it is sufficient that the receiving UE knows at least parts of the signal transmitted by adjacent UEs. Pilots, such as CRS, PRS in 4G, which are multiplexed in a UE's transmit signal stream anyway, can be used for that purpose. Another option is to transmit dedicated ranging signals, which are multiplexed into the UE's transmit signal stream from time to time. However, there is no need to establish mutual connections between UEs.

Figure 10a and **b** shows examples for cooperative positioning in indoor environments. In such areas, we find a lot of "things" that will be connected. Such devices are, e.g., smartphones, laptops, WLAN, or 5G access points (APs), but also consumer electronics like smart TVs or even home appliances like fridges, dishwashers, washing machines, etc. Many of these "things" of the internet are stationary as shown in the example in **Figure 10a**. Still, their

(a)   (b)

Figure 10. Cooperative positioning indoors. The image on the right is licensed under CC BY-SA 3.0. It is built upon https://commons.wikimedia.org/wiki/File:Mirdif_City_Centre_indoor.JPG by Shahroozporia (own work) [CC BY-SA 3.0 (http://creativecommons.org/licenses/by-sa/3.0)], via Wikimedia commons (a) Home area, (b) A shopping mall.

position might be unknown and must be determined similar to the mobile devices. The a priori knowledge that they do not move can be exploited in that context. Highly mobile things of daily use, like glasses, dog, or cat collar, may be equipped with low-cost transmitters in future. Thus, also these items, which are often lost, become traceable with cooperative positioning methods. Signals from outside the home, like BSs, GPS-equipped UEs outside, etc., might also be received under good LOS propagation conditions. These observations extend the mesh of connected devices and allow positioning in a global coordinate system. In shopping mall areas, as shown in **Figure 10b**, the density of mobile communication devices like smartphones is usually high. With a dense mesh of such devices, LOS propagation conditions among adjacent devices are highly probable, providing accurate ranging capabilities. The mesh reaches outdoor areas through devices near entrances or windows. Meshed devices outdoors can use GPS positioning and serve as a kind of anchor for devices' indoors. Also, stationary indoor APs can serve as anchors. Their positions may have to be determined.

5.2. Cooperative position calculation in mobile radio networks

- *Centralized, network centric*: UEs transmit their measured ranges to adjacent UEs to a network positioning entity, which calculates the positions of the UEs and provides the position estimates to the UEs. This needs a protocol for exchanging information between UEs and the positioning entity. The protocol overhead might cause latency for position estimation, which might be negligible for pedestrians.

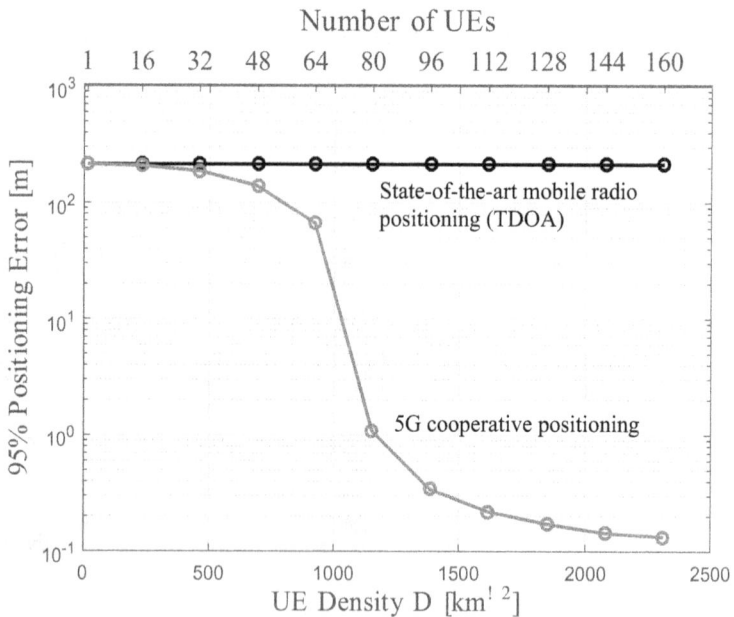

Figure 11. Cooperative vs. non-cooperative positioning performance.

- *Decentralized, UE based*: UEs share their currently estimated state (including uncertainty) to their vicinity. A protocol, which allows broadcasting this information, is needed for that purpose. The state to be shared (broadcasted) includes position and timing estimates, i.e., the offset of the local UE time base to the global system time base. Based on this, individual UEs can estimate their own position locally. This approach allows a "listen-only" mode. A "listen-only" UE does not share its own state estimates, but is still able to calculate a position fix based on the observed signals and state estimates from its neighboring UEs.

Figure 11 shows results about the expectable cooperative positioning performance versus the UE density [21]. UEs have been uniformly distributed in a triangular area between three BSs as shown in **Figure 2**. The simulation parameters are summarized in **Table 2**. For non-cooperative positioning, the UEs' positions are calculated individually. Thus, the positioning error does not depend on the UE density. For cooperative positioning, however, there is a significant performance gain for UE densities in the order of 1000 UEs per km^2 and above. The example shown in **Figure 12** provides a relation to a density of 1000 UEs per km^2, which comprises four sites in a typical urban living area. Each of the living sites may contain devices

Parameter		Value
Carrier frequency	f_c	5 GHz
Base station TX power	P_{BS}	30 dBm
Base station TX signal bandwidth	B_{BS}	5 MHz, uniform power spectrum density
Mobile terminal TX power	P_{UE}	20 dBm
Mobile terminal TX signal bandwidth	B_{UE}	1 MHz, uniform power spectrum density
Noise power spectral density	N_0	$N_0 = k_B T$
Boltzmann constant	k_B	1.381×10^{-23} J/K
Noise temperature	T	300 K
Propagation model BS-UE		WINNER C2 Typical Urban, additional ranging error of 150 m if link is in NLOS condition
Propagation model D2D		free space, communication range is limited to $r_{com} = 50$ m
Base station distance	d_{BS}	400 m
Number of UEs	N_{UE}	1, ..., 160

Table 2. System parameters for cooperative positioning simulations.

Figure 12. 1000 UEs per km^2 means 1 UE per 1000 m^2.

as shown in **Figure 10a**. In shopping malls, as shown in **Figure 10b**, user densities are usually much higher. 5G envisages device densities of 10^6 per km^2 or 1 UE per m^2.

6. Conclusions

5G is envisaged to support a variety of use cases and therefore needs to support precise positioning in many cases. With higher carrier bandwidth, TOA and related measurements can be done precisely. Cooperative positioning will benefit from the dense network and D2D communications. All these will contribute to high-accuracy positioning. In this chapter, we overviewed positioning requirements for wireless communications and the relevant radio-based positioning techniques. Then, we discussed the CRLB and ZZLB. With a simple ML-based adaptive threshold method, the first path of the radio signal can be detected with high accuracy for many wireless channels, especially when LOS is strong. In massively connected IoT, cooperative positioning will provide a further way for precise radio-based positioning.

Acknowledgements

Part of this work has been performed in the framework of the Horizon 2020 project 5GCAR (ICT-761510) receiving funds from the European Union. The authors would like to acknowledge the contributions of their colleagues, although the views expressed in this contribution are those of the authors and do not necessarily represent the project or company.

Author details

Wen Xu[1]*, Armin Dammann[2] and Tobias Laas[1,3]

*Address all correspondence to: wen.xu@ieee.org

1 Huawei Technologies Duesseldorf GmbH—European Research Center, Munich, Germany

2 Institute of Communications and Navigation, German Aerospace Center (DLR), Wessling, Germany

3 Department of Electrical and Computer Engineering, Germany and Technical University of Munich, Munich, Germany

References

[1] 3GPP TS 25.305. UMTS—Stage 2 Functional Specification of UE Positioning in UTRAN. V14.0.0. 2017

[2] 3GPP TS 36.211. E-UTRA—Physical Channels and Modulation (Rel. 14). V14.4.0. 2017

[3] WINNER II Deliverable D1.1.2. WINNER II Channel Models. 2007. Available from: http://www.ist-winner.org/deliverables.html

[4] 3GPP TS 36.305. E-UTRAN—Stage 2 Functional Specification of UE Positioning in E-UTRAN (Rel. 13). V13.0.0. 2015

[5] 3GPP TS 36.355. E-UTRA—LTE Positioning Protocol (LPP) (Rel. 13). V13.3.0. 2016

[6] 3GPP TS 36.455. E-UTRA—LTE Positioning Protocol A (LPPa) (Rel. 13). V13.1.0. 2016

[7] FCC 15-9. Wireless E911 Location Accuracy Requirements. 2015

[8] Fischer S. Observed Time Difference of Arrival (OTDOA) Positioning in 3GPP LTE. Qualcomm Technologies, Inc. 2014. Available from: https://www.qualcomm.com/media/documents/files/otdoa-positioning-in-3gpp-lte.pdf

[9] FCC 99-245. Third Report and Order. Federal Communications Commission (FCC). 1999

[10] Wymeersch H, Seco-Granados G, Destino G, et al. 5G mmWave positioning for vehicular networks. IEEE Wireless Communications. 2017;**24**(6):80-86

[11] 3GPP TS 38.101-1. NR; UE Radio Transmission and Reception; Part 1: Range 1 Standalone (Rel. 15). V15.0.0. 2017

[12] 3GPP TS 36.101. E-UTRA—UE Radio Transmission and Reception (Rel. 14). V14.5.0. 2017

[13] Xu W, Huang M, Zhu C, Dammann A. Maximum likelihood TOA and OTDOA estimation with first arriving path detection for 3GPP LTE system. Transactions on Emerging Telecommunications Technologies. 2016;**27**(3):339-356. DOI: 10.1002/ett.2871

[14] Kay SM. Fundamentals of Statistical Signal Processing: Estimation Theory (Vol. 1 of Prentice-Hall Signal Processing Series). Upper Saddle River, New Jersey, USA: Prentice Hall PTR; 1993

[15] Lin X et al. Positioning for the internet of things: A 3GPP perspective. IEEE Communications Magazine. 2017;**55**(12):179-185. DOI: 10.1109/MCOM.2017.1700269

[16] del Peral-Rosado JA et al. Impact of frequency-hopping NB-IoT positioning in 4G and future 5G networks. In: Proceedings of the 5th IEEE ICC Workshop on Advances in Network Localization and Navigation (ANLN); 2017. DOI: 10.1109/ICCW.2017.7962759

[17] Xu W et al. Techniques for determining localization of a mobile device. Patent application PCT/EP2017/070798

[18] Raulefs R, Dammann A, Jost T, et al. The 5G localization waveform. In: Proceedings of the ETSI Workshop on Future Radio Technologies: Air Interfaces; 2016

[19] Ziv J, Zakai M. Some lower bounds on signal parameter estimation. IEEE Transactions on Information Theory. 1969;**15**(3):386-391. DOI: 10.1109/TIT.1969.1054301

[20] Musso C, Ovarlez JP. Improvement of the Ziv-Zakai lower bound for time delay estimation. In: Proceedings of the 15th European Signal Processing Conference (EUSIPCO); 2007

[21] Dammann A, Raulefs R, Zhang S. On prospects of positioning in 5G. In: Proceedings of the 2015 IEEE International Conference on Communication Workshop (ICCW); 2015. DOI: 10.1109/ICCW.2015.7247342

www.ingramcontent.com/pod-product-compliance
Lightning Source LLC
Chambersburg PA
CBHW081240190326
41458CB00016B/5855